# Pulling Wings From Butterflies

## - Tercio De Varas -

By Capt. Mike Simkins

Published in 2020 by Attica Publishing Ltd.



www.pullingwingsfrombutterflies.com



Attica Publishing Ltd.

71-75 Shelton Street,

Covent Garden,

London, WC2H 9JQ

www.atticapublishing.com

Ordering Information:

Quantity sales. Special discounts are available on quantity purchases by corporations, associations, and others. For details, contact the publisher at the address above.

Orders by U.S. and U.K. trade bookstores and wholesalers. Please contact

info@atticapublishing.com

Printed in the United Kingdom



ISBN: 978-1-8383130-1-2

This book is dedicated to all those who walk towards the flames
when everyone else claims there is no fire.

To the memory of my father Alan Simkins
"Remember. It will always be the pilot's fault."

## *Foreword*

by Justice George W. King and Mark Lanegan.

### *Justice George W. King*

*"...from the safest places, come the bravest words..."*
New Dark Age - The Sound - 1981

In 1986, as a 30 year-old indie music-obsessed employment law lawyer, I became a devotee of the Manchester post-punk outfit B.F.G. (Don't scratch your head as you likely haven't heard of them, until now.) After a clutch of irrepressible singles ("Amelia/ Paris" and "Western Sky") and two brilliant albums, ("Fathoms and Blue") B.F.G. disappeared off the radar screen, much to my dismay. Fast forward many years, and in my quest to solve the mystery of a long-promised third LP, I got to know Mike Simkins – first via the internet and eventually in person.

Upon learning of my career in law, Mike was eager to share his story. And what a story it was. From the early "punk" days as Mike Simii to a "proper job" (as he described it) as Captain Mike Simkins, commercial pilot for a globally-known airline. I learned that he had been severely disciplined and, just before we connected, had completed a legal struggle of epic proportions.

I immediately thought of the verse from The Sound's "New Dark Age," another British post-punk band, when Captain Mike asked me to write this introduction to "Pulling Wings From Butterflies", his tale of the epic legal struggle against his former employer to advocate for the safety of commercial airline passengers.

While we may not express it as lyrically as The Sound's Adrian Borland, from the days of dealing with a schoolyard bully through adulthood, we have all encountered (and continue to encounter) situations where the expression "from the safest places come the bravest words" ring true.

I think it is safe to say that most persons harbour dislike for those speaking brave words from safe places.

Captain Mike Simkins is not one of those people. His story is the very antithesis of that cowardly behaviour—Mike had the courage and conviction to speak his "bravest words" from a position that no-one would dare suggest was his "safest place". In fact, he uttered those words from what was an unsafe place for himself and his career. But more importantly to him, he refused to fly because he believed that could have resulted in his passengers being guided to a most unsafe place.

Perhaps the lyrics from a B.F.G. song "Another Island" from the LP Fathoms foreshadowed what was to come:

"With heavy heart we hit the ground
We fly so high to reach so low
The busy streets of empty faces outside
The distance I feel begins to grow."

When asked to fly his charter plane on a daily schedule Captain Mike believed was beyond the legal hours of duty safety limit, he took a stand. He did this because perceived it was a matter of safety-the safety of his crew, his passengers, the public, and lastly, himself. He paid the price for those brave words as an airline pilot with the potential loss of his reputation as an aviator.

This is a story about redemption and one person's struggle to challenge the consequences of those bravest words.

Along the way, you will discover the world of post-punk music and the economic brick wall that most artists crash into at some point. In the case of the now reformed B.F.G., Mike Simkins was young enough when his first career in music abruptly ended to pursue a more traditional career path as a pilot. That plan ultimately led to this story.

As a jurist, I can't and won't comment on the legalities or issues regarding the legal case you are about to discover. That would not be appropriate. Accordingly, what I can tell you as a reader, is that the story which ensues is a compelling legal case study with a narrative that is as rich in complexity as it is illuminating, with Captain Mike Simkins laying bare the prescient issues of airline safety we all ought to be aware of.

In summation, I say, come along for the gothic dance music that was (and is) B.F.G., but fasten your seat-belt for a hair-raising story of a legal case with the turbulence of an airplane ride into, and out of, a hurricane.

Bravest words?

Indeed.

***

*Mark Lanegan*

Captain Mike Simkins has lived an extraordinary life and has an extraordinary story to tell. In the early eighties through the early nineties he was known as Mike Simii, the hugely talented musician who first came up playing drums in the band Lavolta Lakota who were signed to Manchester's fabled Factory Records, and in the mid-eighties grabbed the reigns and started his own band, writing the music for, and recording the records of legendary post-punk/Goth/Darkwave band, B.F.G. Peers of such groups as New Order, The Membranes, Section 25, and Phantasmagoria-era Damned, B.F.G. hit instant cult status with their very first 12 inch single, 'Paris/Amelia,' to this day a highly sought after piece of vinyl, considered a Goth classic. Later ep 'Western Sky' and single 'Higher Than Heaven' brought them further acclaim, with Melody Maker's Mick Mercer declaring 'Higher Than Heaven' the greatest indie single of the year 1987. B.F.G. called it a day in 1991 and, in a complete 180 degree turn, Mike's already incredible life took a jump into a world so unexpected, it reads like something out of a Hollywood movie, and he began a second act as a commercial airline pilot, which is what this amazing book is really about. In 2019 my band was thrilled when John Robb's Membranes agreed to come on tour with us and I was shocked to realize it was Mike Simii behind the drum kit, playing with as much grace and power as when I first heard him in my early twenties. That's where Mike Simkin's life had begun to come full circle and this uniquely talented, multi-faceted, generous and big-hearted man was back in the place I had known him for. To find out what he has experienced, where he has been and what he has done in the last couple of decades is as compelling a story as you'll ever read, and every word of it is true.

# CONTENTS

# *Introduction*

*"Rules about maximum FDP [Flight Duty period] and not flying whilst fatigued are made with the safety of crew and passengers in mind. If they are not complied with, there is an obvious increased risk to those flying."*

Judge Slater

I made the difficult decision to no longer fly as an airline commander for one reason - trust, or rather the lack of it.

Before an aircraft departs, a commander must trust that the first officer is vetted by the airline, is suitably qualified and has the appropriate experience. They must trust that their colleague on the flight deck is well trained, psychologically stable, and will respond appropriately in an emergency. A commander must trust that any safety task delegated to cabin crew will be performed with the utmost professionalism. He must also trust that the team will have the confidence to report any issues that may concern them. When signing for an aircraft, a captain must trust that the engineers have completed their checks correctly and that the aeroplane is in a fit and proper state to depart the aerodrome safely.

With very few exceptions, for twenty-five years, thirteen thousand flight hours and over four million miles, I trusted all the above with my life. If you were a passenger, so did you.

Along with trusting the above individuals, it is also vital that a commander has complete trust in the honesty and integrity of airline managers, directors, support staff and their legal team. They must trust the honesty and integrity of the regulator who oversees and enforces the legislation designed to protect passengers and crew members. They must trust that the regulator will comply with its own published prosecution policy in the event of transgression.

A pilot must trust that the union will provide legal support to any crew who raise serious safety concerns when refusing to place commercial interests ahead of passenger safety.

Without that trust, a captain's position becomes untenable.

I no longer fly; my position became untenable.

I no longer trust.

Pulling Wings From Butterflies explains and exposes the culture within the aviation establishment that has allowed such a situation to develop where it appears, some airlines, are able to commit criminal acts that directly affect passenger and crew safety with apparent impunity. Ultimately this culture resulted in my very public legal case against the airline Thomas Cook. Like an iceberg, however, only ten per cent of the real story has so far been revealed.

Over the series of three books, I will refer to documents and experiences from my case. I will explore and explain other real-world examples where safety was *not* the number one priority. My ultimate aim is to expose the dark underbelly of an airline industry, both in the UK and USA, that publicly profess to always 'place safety first'.

The reality is, it appears to be an industry infected by 'Regulatory Capture', 'Tombstone Safety' and the 'Normalisation of Deviance'.

Without these crimes of governance, all destructive and dangerous dark forces, the current Boeing 737 Max fiasco would never have taken place. Why else would a captain refuse to report to the CAA the details of a flight duty that was so fatiguing it resulted in two of the three pilots asleep at 4000 feet on final approach? Why else would crew fail to report to the airline or regulator a situation when all three pilots fell asleep over the Atlantic?

What happened to me is not unique, apart from the fact that I said NO!

It explores how two previous serious safety issues I had predicted would result in tragedy if not immediately attended to, were not appropriately dealt with by the regulator, the CAA. Both subsequently resulted in fatal air crashes. Haunted by these failures, the underlying driving force at the heart of my story is my effort to prevent a third.

Since my case hit the worldwide headlines, I have been inundated with requests for advice from pilots facing similar issues to those I have dealt with. Stories of pilot unions turning away and regulators apparently looking after the commercial interests of the airlines rather than passengers and crew are aplenty. In short, a vast swathe of the aircrew population has lost faith in those very organisations designed to protect passengers and crews interests. This is the reason I decided to write the series 'Pulling Wings From Butterflies'.

The three volumes are named after the three acts of the barbaric tradition of bullfighting.

'Tercio de Varas' signifies the act where the animals' reactions and condition are assessed by goading the creature. 'Tercio de Banderillas', is the period where three pairs of banderillas are thrust into the bull's shoulders to weaken the beast. 'Tercio de Muerte' is the final act where there will ultimately be one victor. In the world of aviation, this should be passenger and crew safety. Through three volumes, Pulling Wings From Butterflies actively questions that assumption.

The books serve as a rallying call for those disenfranchised by the apparent abject failure of individuals and organisations that 'claim' to represent passengers and crews interests. The series also provides real, *need to know* information that companies, regulators and even unions seem not to want in the public domain. The law is there to protect employees. Unfortunately, most are unaware of the safeguards available. I aim to reveal those safeguards.

'Pulling Wings From Butterflies' is the true story of my standing alone against the might of the entire aviation establishment when I refused to place corporate profit before passenger and crew safety. It also charts the eventual, predictable demise of one of the UK's most recognisable companies due to appalling mismanagement after 179 years of trading. The journey begins with my performing as a rock drummer throughout Europe with some of the biggest names in the music business, to ultimately becoming a commander on one of the largest and most technically advanced aircraft in the world. It ends when I finally found where true fulfilment lay; when life came full circle with a thirty date European tour with one of rock music's most iconic figures.

The details within are based on documents provided to a judge. They are also based on audio recordings, notes, and minutes made during various meetings and investigations. My conclusions are my honest opinion based on documented evidence and facts. I, therefore, invite you to form your own conclusions as to the integrity, competence and honesty of the individuals and organisations described, based on the facts as the case unfolds.

Due to the abject failure of the aviation establishment and government to address the grave issues contained within Pulling Wings From Butterflies, the details are released in the public interest.

# CHAPTER ONE

## *Erode The Safety Margin*

*"If I were to remain silent, I'd be guilty of complicity."*

Albert Einstein.

Clad in a distressed leather jacket, and the only one in the room who needed to have stood closer to the razor that morning, was Eddie. Pencil and notebook in hand, every scribbled sentence forcefully punctuated. The suits surrounded Eddie, whose choice of attire hinted at his chosen profession.

The closest of the suits was Scadeng. He was sweating like Nixon.

I finally had my opportunity: the moment I had been awaiting for a long and exhausting, eighteen months.

It was the last day in a week-long legal case going head to head against the might of a multibillion-dollar airline.

Representing the airline was a well-respected barrister supported by a legion of Thomas Cook legal and HR advisors. Representing me was a fifty-four-year-old Doc Martens-wearing airline captain who had spent most of his twenties sitting in the back of a transit van touring Europe as a drummer in a loud, post-punk, rock band. Unfortunately, sitting in front of a judge, my mix of experience from paradiddles to pre-flight checklists seemed to count for nothing in this formal and unforgiving environment. At this precise moment the most useful skillset would have been forged, not in a cockpit or sitting behind an oak custom drum kit, but in legal chambers.

The suit sitting next to me, Rob Hunter, was the head of flight safety from the pilots union BALPA. Rob was one of the leading fatigue and medical experts within the field of aviation. However, his appropriate legal knowledge was even more limited than my own.

The judge had just called 'seconds out' and had invited me to continue my cross-examination of the senior director of the airline. I had operated as captain for this airline for the last sixteen years, faultlessly. Sitting in the witness stand was Paul Hutchings, the airline's director of flight operations (DFO). In front of the DFO, the judge and myself were lever arch files. Each file contained over 1000 pages compiled for my case. Page 963 contained the paragraph I needed to nudge, provoke, irritate and generally manoeuvre Hutchings towards. Page 963 was where I wanted him. Page 963 was the 'Tercio de Varas.'

This case was not just about myself. It was not even about my immediate working colleagues. Ultimately the judge's decision would resonate within the entire pilot and flight crew community throughout the country, possibly even the world. It would have a direct effect on the safety of the millions of passengers who entrust pilots and airlines with their families' lives every day of the week.

Where I was sitting there should have been a knowledgeable, experienced barrister able to interrogate and cross-examine any witness who was obstructive or evasive. Where the doctor was sitting there should have been a well-briefed solicitor taking notes and offering support to the superior advocacy skills of a legal professional.

As a youngster, I had played a lot of chess, and this whole case had felt like a multi-layered 3D game played blindfolded and without knowing the rules. For the past four days, I had stumbled while finding my legal feet, working out how all the individual pieces moved. Somehow I was still in the game. Somehow I still had all of my pieces and, at least so far, I had been able to cover any attack by the opposition. Even more amazingly, I had managed to position my metaphorical rook on the opponents back row, and I was now picking off the Thomas Cook pieces in rank order. It felt like I was currently controlling the game, but I knew there was still a long way to go.

Before I continued my cross-examination of Hutchings, I consciously waited a minute or so. Let him contemplate what was about to happen. My pen idly traced the lines of some random text as if deep in thought. The reality was, this was more about setting a scene than absorbing any further information. My thinking had already been done.

Each second that passed would amplify a simple fact that Hutchings had to accept, appreciate and consider. As the DFO of a major airline, he was about to be cross-examined in front of a judge, under oath, witnessed by two lay legal experts by one of his most senior aircraft commanders.

I let him marinate for a few moments longer in the knowledge that, unlike the previous eighteen months, he would now have to answer *my* questions. No longer, could he say something and then claim he had not said it. No longer could he instruct his subordinates to fire the figurative bullet and claim he knew nothing about the attempted drive-by. No! This time it was me, him and a judge. This time he knew everything that he said throughout my cross-examination would be recorded, analysed, countered and ultimately, judged.

I leant forward and stared at an individual who I should have been working alongside as part of a team. Instead, I was now intent on exposing and then demolishing Hutchings integrity. I held his gaze for a few moments as he shifted uneasily in his chair.

'In your capacity as the director of flight operations for Thomas Cook, can you confirm you are content for your passengers to be flown by pilots that are operating at a level of fatigue that results in the equivalent effectiveness of someone at the drink-drive limit?'

Thomas Cook's barrister sat bolt upright. 'Objection Ma'am!'

The judge looked at me with a wry smile. 'Sustained! Captain Simkins, It is not permissible to ask a leading question.'

I smiled back. 'I apologise ma'am, I am a pilot, not a barrister.'

I could blame my lack of legal etiquette and knowledge for my faux pas and the resulting light wrap on the knuckles by the judge. The point was, I had now communicated to the DFO but, more importantly, the tribunal, the central premise to the last five days of legal argument.

It was a simple question. Is it acceptable for an airline to take action, up to, and including dismissal, against a highly experienced airline captain because he refused to fly an allocated duty whilst so exhausted it would have had serious safety implications to his crew and passengers?

The judge then offered to ask my original question in a more legally appropriate form. I had learned quickly. Being litigant in person did have certain advantages.

The DFO looked uneasy. His usual self-assured demeanour had evaporated quicker than Prince Andrew's sweat. For the first time, he was now *not* playing on home territory. He was no longer controlling the game and no longer calling the shots. For the last few days, he had witnessed my cross-examination of most of his flight ops management team. Now it was his turn to answer. It was *his* turn in front of the wall.

My original question and the subsequent rephrasing by the judge rested on the significance of a printout provided by the company that predicted a pilot's level of fatigue based on previous duties and sleep patterns. As I had planned, the DFO's answer to my now legally acceptable question left the door wide open for the only medically trained person to have looked at the printout to throw the knockout punch.

Had Hutchings been outmanoeuvred? Was his vision of a united front with his well-disciplined Thomas Cook management disciples, lined up against a single upstart, starting to resemble a house of cards?

Hutchings' own crewing manager, Jane Marsden, already had inadvertently exposed his and the company's view about the risks the airline appeared willing to take with passenger safety. She had little idea of the significance of her comment when she made the unguarded statement: "We cannot avoid operating duties in this zone (levels of effectiveness below the dotted line, i.e. below 77%) due to the nature of the program that we operate."

Rob performed the 'tercio de banderillas'. 'Mr Hutchings, please turn to page 545 in the bundle.' Hutchings began to hunt for the elusive page that the doctor and I already had open on the desk. 'Do you think the level of fatigue, where the trace is in the yellow zone, may endanger the flight?'

Hutchings' answer stunned not just myself, but the entire public gallery. 'This run of duties was not that unusual' before adding 'but I accept that that it could erode the safety margin.'

I looked over at the airline's solicitor, Lindsey Murray. She could not prevent herself from grimacing. This was not good. This was not good at all for a major company that was a household name.

I wrote it down in my notes "Erode The Safety Margin." I boxed it, underlined it, then stared at it and let the full implications of that statement sink in, as I suspected, did the judge and lay members of the tribunal. They were still all wide awake.

It was now time for me to administer the 'Tercio de Muerte'.

'Paul, turn to page 545.' I continued, 'You can see the trace shows that my predicted level of effectiveness for the duty I refused was down at 70%.'

The reply from the DFO produced yet another jaw-dropping moment. 'Yes, but if you look, that is only towards the end of the flight.'

I looked towards the judge, only to find that she, along with both lay members, was busy scribbling down the latest DFO soundbite. The most critical stage of any flight when any mistakes are significantly amplified is the approach and landing phase. Yet here was the DFO of one of the UK's largest airlines attempting to play down the significance of an '*effectiveness*' of seventy per cent because it was only at the '*end of the flight*'.

I delayed the next line; like a child slowly opening a Christmas present, I wanted to savour every second.

'Please turn to page 963 of the bundle, Paul.'

Again the DFO complied. The text came straight from the manual for the software that had produced the printout— the printout upon which Thomas Cook was basing their entire case. I slowly read the text for the benefit of the tribunal.

'In order to provide some guideposts to relate levels of effectiveness to elevations of risk of an error or accident, consider the following. Considering speed of reacting to a signal, effectiveness as low as 70% lengthens reaction time about as much as being legally drunk (0.08 blood alcohol content).'

I hammered this final nail into his and the airline's position. '0.08 blood alcohol content is the equivalent to the drink-drive limit in the UK or four times the blood alcohol content limit for flying.'

'I admit it's not ideal' came the less than ideal, at least from the airline's perspective, reply from Hutchings.

We had already heard from the flight ops management team that my run of duties was by *no means unusual*. So how many more pilots are flying at the effectiveness of that of a drunk driver?

Time to close this out;

'Is it *your* decision or *mine* if I am or if I am not fatigued?'

'Your decision.' Hutchings took a deep breath to launch, yet again, into another long-winded monologue.

I cut him short. I had what I had needed. 'No further questions ma'am.'

Eddie continued to scribble, Scadeng continued to sweat.

# CHAPTER TWO

## *The Music Years*

*"There is only one person who does more flying than me... And that's the pilots, and I hate it."*

Andrew Gold.

What possible twists and turns in life could ever have brought me here? Currently, I'm sitting on a tour bus, having just embarked on a thirty-date European tour, drumming for one of the most influential post-punk bands of all time, all whilst writing about corruption and collusion within the aviation industry?

As I complete each page on a beaten up Mac Book Pro, I take a moment to contemplate this bizarre set of circumstances as the tour bus makes its way between Berlin and Copenhagen. The route between the two cities being no less circuitous than my journey from rock drummer in the 80s to operating as a commander on one of the most advanced aircraft in the world, to then confronting the entire aviation establishment.

But now, here I am, about to drum for the first time in thirty years, in front of an audience of around 1,500 a night for a band who were a significant influence on many artists including Nirvana's Kurt Cobain and Queens of the Stone Age's and Screaming Trees vocalist Mark Lanegan.

If you discount the first sixteen years of my life, this journey began in 1977. It was the heyday of punk music. My most pressing concerns were which brand of medication would cure acne and would sugar water have enough strength to hold my hair in the same style as John Lydon's?

As a spotty, spiky-haired youth, I was now sitting in Ashton

Grammar's sixth- form careers advisor's office. I had been asked what my preferred choice of occupation was.

'Airline Pilot.'

With that, the teacher-come self-professed employment expert rocked back on his chair, took a drag on his pipe, and stared back at the adolescent sat in front of him. His verdict, delivered with an almost un-contained relish was that I did not have a hope in hell. He informed me that during his time as a career advisor, hundreds had sat where I was now sitting. Many had pronounced that they wanted to be a pilot, yet only one had ever made it as far as even being selected as a cadet. Even then, they had only lasted two months at Hamble, the British Airways Training College, before they were 'chopped' as not up to the task. He asked if I had any other ideas.

The lifting inflexion at the end of my next answer for the next stab at a fulfilling career choice probably gave the game away as to my inner lack of belief if this would be possible.

'Drummer.'

'Can you drum?'

My response in the negative did not even warrant an aural response from the nicotine impregnated oracle. The shaking of his head was the only outward sign of any acknowledgement to my answer.

There I was, a sixteen year old, with middling school reports with dreams of the unachievable.

The old cotton mill town of Ashton-under-Lyne did not produce the likes of airline pilots or rock stars. The local comprehensive school where I had spent the first five years appeared to have only two goals. The first, to keep as many yobs off the street as it could for a few hours a day. The second was to train the legions of factory workers that would ultimately operate in some capacity or other in the numerous 'dark satanic' cotton mills dotted around the Lancashire town. There are still a few mills standing that managed to survive the 80's mill cull by celebrity demolisher Fred 'did ya like that?' Dibner.

Somehow I had managed to gain a brace of 'O' Levels after five years at the local comprehensive. A grade 'A' in physics and 'B' in metalwork were the highlights. A position making telescopes, however, never appeared over the horizon.

These academic highs were enough to secure a place at the local grammar school sixth-form.

It was only once I had left the comprehensive and began studying for A levels, did I finally feel my classmates were individuals I could relate to. Unfortunately, this caused the study to take a backseat to a social life that I had been deprived of for the previous five years. Conversations would centre around the embryonic punk music scene in Manchester, a scene very much championed by a newsreader on Granada, the local TV station. His name was Tony Wilson.

Wilson had managed to sweet-talk the studio owners at Granada into giving him his own music show entitled 'So It Goes'. The reality was, it was the only music show worth watching. It was not the usual dire musical offerings, performed by vapid, grinning, airbrushed puppets, the strings of which were held by some self-appointed musical impresario. This music was produced by young, mostly angry, mostly ugly individuals who had something to say and by God were they going to say it. It was exciting, it was new, but there was more to it than that. *Much* more! It was the mindset of many of those involved that was the biggest attraction. Bands who appeared on Tony Wilson's show included The Buzzcocks, The Sex Pistols, Joy Division, Siouxsie and the Banshees, The Stranglers and The Clash. The groups were filmed playing live in Manchester, with a local poet, John Cooper Clarke, acting as a cleanser of the musical pallet. He was a punk sorbet, mixing social commentary with laser-like humour. Some forty-odd years later, it is wonderful to see John finally get the recognition he has, for so long, deserved.

Along with 'So It Goes', my musical education was supplemented every weeknight from 10:00 pm. The strains of 'Picking The Blues' by Grinderswitch would introduce what many consider to be the most influential radio show to have ever been hosted on the BBC, the 'John Peel Show'. Peel was right there at the forefront of this musical and social movement that punk represented. Along with the bands who were already appearing on 'So It Goes', Peel would play The Dead Kennedys, Talking Heads and Television from the USA. Stiff Little Fingers and the Undertones from Northern Ireland, The Adverts, The Jam along with his all-time favourite band, The Fall.

Stiff Little Fingers were, and still are, fronted by Jake Burns. They were a noisy, poorly- produced punk band that had formed at the height of the troubles in Northern Ireland. During the late '70s, early '80s not a day would go by without news of the IRA murdering a Protestant or member of the police or the UDA killing Catholics. There were also the constant bombings being carried out in Ireland and on the UK mainland. Despite this, Stiff Little Fingers sang about precisely what it was like to grow up surrounded by that level of turmoil and strife. They made statements in those tracks that many individuals had been shot for saying less, on both sides of the Irish political divide.

These young lads were brave, they were bold, and they were utterly brilliant. They still are. The band's debut album was a history lesson in how it was to be young and defiant when two opposing factions attempting to convince you that their ideas and beliefs were the only right way, the only truth. The factions' methods of persuasion extended to violence, torture and murder.

If I wasn't listening to John Peel's show, I would be sitting upstairs listening on my air-band radio to aircraft inbound to Manchester Airport. I was fascinated and intrigued by the radio exchange between aircraft and ATC (Air Traffic Control);

'Established on the localiser.'

The approach controller would reply;

'descend further with the glide, contact tower on 118.625.'

It all sounded so very technical, so very cool. Alternatively, I was working out which way round to hold a guitar in a feeble attempt to play Bobby Shafto from my 'Tune a Day' guitar book. Given the pace of my improvement, a more apt title would have been a 'Tune a Millennium'.

The first record I ever owned was by The Sex Pistols, the band the establishment loved to hate. They not only recorded one of *the* greatest rock and roll albums of all time but their frontman, John Lydon, was not afraid to say things as they were, even if it was not what the establishment wanted to hear. Maybe that's why they hated him.

Lydon was asked by a BBC reporter, no doubt in an attempt to make him look foolish, who would he kill if he could.

'Jimmy Savile', Lydon replied immediately. 'We all know what he's up to, but nobody is allowed to say anything about it.'

That one comment resulted in the BBC immediately banning any Sex Pistols tracks being played, or even being referred to on the nation's most listened to station.

As anyone who lives in the UK is now aware, it transpired Savile was a prolific, predatory, paedophile. Many at the BBC knew this back in the '70s, but the BBC loved Jimmy.

It later descended into a farce when the Sex Pistols track 'God Save the Queen' reached number one. The BBC redacted the band's name and the title of the record as if it simply did not exist. It might not have existed in the eyes of the BBC, but for many, it opened ours.

The important issue here, for daring to challenge a high profile member of the broadcasting establishment, Lydon and his band were blacklisted by the only broadcaster who could breathe life into any music career on a nationwide basis. A brave or stupid act of raising an issue that needed to be raised? The fact was, a spotty youth with an aversion to a toothbrush had.

So why would the establishment attempt to suppress the punk movement's figurehead when he spoke out? Surely he should have been applauded. Instead, he was shunned. How many other children suffered because a singer in a band the establishment loved to hate had been silenced? Had they something to hide? Was someone getting too close to the story? The BBC establishment realised someone was going to upset the apple cart and derail that gravy train that just kept on giving to the chosen few.

This was an example of an organisation, in this case, the BBC, turning a blind eye to criminal activity simply because Savile, at the time, was one of its biggest stars. He made the BBC money. Also guilty of looking away were the many charities that Savile had financially supported. What were a few lives ruined if Savile continued to bring in a few million?

Anyone who dared rock the establishment's boat could not be seen to get away with it for fear of others voicing their valid concerns. As for Lydon, he promptly formed the band 'Public Image Ltd' along with ex Clash guitarist Keith Levene and Jah Wobble.

So here I was, sixteen with a love of music but not able to play any instrument and a dream of becoming a pilot without the necessary academic abilities needed. The future did not look too bright.

After my meeting with the 'Yoda' of career advisors, the college had held a 'next step' seminar. The concept was to give suggestions and advice on which courses were open after completing 'A' Levels. British Airways were conspicuous by their absence. There was, however, a very smartly dressed chap. His boots were so reflective it was a pity the engineers responsible for the mirrors in the Hubble Telescope had not taken lessons from him on how to polish.

If I didn't stand a chance of getting into Hamble as a British Airways cadet, maybe I could see if Her Majesty might pay for my flight training if I joined the RAF.

There was, however, a catch.

The shiny-booted officer informed me that because it cost upwards of a million pounds to train each fighter pilot, I would have to serve queen and country for at least twelve years. I would be an ancient man of thirty by the time I had paid my dues! However, there was an even bigger issue and one that would be impossible to resolve.

The aforementioned Stiff Little Fingers debut single was a track entitled 'Suspect Device'. It contained the line, 'question everything you're told'. This was precisely the mentality that would be drummed out of any aspiring fast jet pilot in the RAF. You were there to follow orders, nothing more, nothing less. When the commanding officer orders you to jump, you jump. The only question allowed to be asked? 'When can I land?'

My analysis of the predicament I faced as a sixteen-year-old was confirmed as correct many years later. A TV documentary followed the highs and lows of RAF pilots. During one episode, a pilot was challenged.

'If your Commanding Officer told you to bomb a village, would you do it?'

After several minutes of side-slipping the question, the pilot finally admitted;

'Well, I suppose I would have to if I were ordered to.'

No, as far as the RAF was concerned, I was certainly not made of the right stuff. Maybe it was just that I thought about things too deeply. Perhaps it was just that I thought about things.

With the dream job being entirely out of reach, upon leaving Sixth Form, I joined the Nat West Bank. Secure employment, cheap mortgage, nine to five working hours with every weekend off. What was not to love about it? Well, pretty much everything. I hated it.

As my record collection began to grow, so did the realisation that life had a lot more to offer than printing cheque-books. I needed to be part of this seismic musical and cultural shift. The only problem was I had not advanced from playing 'Bobby Shafto' on the acoustic guitar. Before punk came along, this inability to play any instrument would present something of a hindrance. With punk, it was not only positively encouraged but also seen as an advantage. Unfortunately, my guitar playing at this stage would have stretched even the boundaries of the punk ethos. There must be another instrument I could play without the limiting restriction of actually having to learn to play it.

My father and I would debate anything, life, music, careers. Some might say, argue. It was definitely debate. Whatever position on any subject one of us would take, the other would present the opposing view.

He knew I was, by now, considering getting a set of drums. They seemed easy to play, just bash about with no need to learn or understand such complicated items as notes and chords. Looking back, I might have given the game away that I was serious in my intentions. With the number of brochures from the likes of Premier, Tama, Pearl and Ludwig regularly dropping through the letterbox, I was not exactly concealing my goal to make a racket.

My father was not ambiguous about his view of his son's musical instrument of choice. 'Not while you're living in this house, lad.' He pointed out that we already had a piano and a battered acoustic guitar.

That was my life mapped out, a career in the bank with no escape. Buy a house on an estate with a cheap mortgage and be shackled for life. There was, of course, crown green bowling to look forward to in my retirement, but there again, I have never looked good in white and a flat cap.

My parents were due to go on holiday, and it was while they were away, the phone rang. It was my friend Keith. Unlike myself, he had managed to secure good grades at 'A' level. He was now studying for a degree at Liverpool University. Keith informed me that Stiff Little Fingers were due to perform at the University. Perfect timing! It was the day after my twentieth birthday, and I could go to the concert and be back, ready for the next stupefying day at the bank.

We arrived at the venue early, the hall already full. Before we knew it, the band were on, and, as one, the crowd began jumping up and down as SLF launched immediately into 'Gotta Get Away'. We both quickly realised it was going to be a hot and sweaty couple of hours. The pounding drums and guitars turned up to the max still could not drown out the angry growl from frontman Jake Burns.

As the crowd jumped, to preserve your front teeth, you had no choice but to follow suit. Not only that, you had to ensure you bounced in sync. If you were out of phase with those around you, you risked a head smashing against your chin and removing gratis, your incisors. Within thirty minutes, we were exhausted, running on pure excitement and adrenaline that only a band at the top of their game can induce.

I noticed hands reaching up at the front of the audience. The drummer was throwing drumsticks into the crowd. One came twisting and twirling through the air in my direction. It resembled a line out in a rugby union game, apart from the fact that due to our synchronisation abilities, the crowd, mostly, still had all their own teeth. My arm went up, but the stick eluded my grasp. A scramble ensued around my feet. Eventually, a victorious youth in a leather jacket appeared displaying his spoils. Pumped on adrenaline, I announced to Keith that if I managed to grab a drumstick that evening, I would buy a set of drums that weekend. What were the chances of another one coming my way? If I did walk away from the gig with a stick, I would take it as a sign from John Bonham, the great percussionist in the sky, that this was the route I simply *must* take!

Moments later, another stick headed in my direction. This time I was ready. I launched myself skyward, the arc and spin rate of the stick was perfect, hand and wood docked like a Soyuz with a space station. The next time my feet hit the ground, I was the proud owner of a Pro-Mark drum stick. The previous careless owner had been Stiff Little Fingers' drummer, Jimmy Riley.

That moment changed my life. True to my word the next weekend, my bank account was two hundred pounds lighter. It was a bottom of the range bright red 'Pearl Apollo'. It was love at first sight. Within a few short weeks, I had joined my first band.

For the first couple of years, I juggled working full-time at the bank while playing with various bands from the Manchester and Liverpool scene. I eventually joined a group who had a record deal in 1984. The band were Lavolta Lakota, and they were already signed to the highly regarded Factory Records or, to be more precise, Factory Benelux. This was the label formed by the previously mentioned Granada TV personality, Tony Wilson. Factory Records was also the home of Happy Mondays and Joy Division, with Lavolta Lakota signing to the label a few years after the death of Joy Division's singer Ian Curtis. After his demise, Joy Division morphed into New Order. Peter Hook, the bass player with both bands, had not only produced Lavolta's first record but also mixed the live sound whenever we played. Numerous appearances followed supporting the Manchester legends at some of the UK's biggest venues. Meanwhile, I still had to put on a suit the next morning, ready for sitting behind the counter at the National Westminster Bank!

This period presented a huge learning curve and opened my eyes to the 'business' aspect of the music industry.

My first face to face meeting with Hooky was an introduction to the reality of being in such an iconic, ground-breaking and influential band. A band so important, two films have since been made about him and the rest of his bandmates. The reality is, there was no money in being iconic.

I had been invited down to a bonding session with my new bandmates. The venue was a popular hang out known as the Gallery where a band were appearing from the USA. It was their first UK tour. There was virtually no one there, maybe no more than twelve or so in total. Not long after I walked into the club, Dave, the singer from Lavolta, arrived. With him was Hooky, a man who even now has singularly failed to work out how to shorten his bass strap. At this time, New Order were at the pinnacle of their popularity. 'Blue Monday' was in the charts and numerous TV appearances, world tours and worldwide recognition would ensure the track would become the biggest selling twelve-inch single of all time. Anyone would, quite rightly, assume that Hooky was, therefore, a very well - off individual.

Moments after being introduced, Dave leaned over to ask could I lend the multi-millionaire ten pounds. It transpired Hooky was anything but a multi-millionaire. This fact had resulted in the ATM declining his request for a cash withdrawal. Lesson number one in the music business; fame does not equate to a decent bank balance. It was an industry based on smoke and mirrors.

Years later, I was to discover it was the exact same situation within the aviation industry. An airline will show a handsome pilot walking through the terminal with a bevvy of stunning-looking cabin crew to give the impression of high-end glamour and luxury. When you consider, a ten year BMI cabin crew member's salary was £15,000 per annum, and the cabin crew spent most of their time selling duty-free, cleaning up vomit and pouring tea or coffee, it suddenly does not look so glamorous.

Similarly, a record label will attempt to sell to the public an idea of success, intrigue and glamour. Flash cars. Designer clothes. Bands hanging out with the current crop of the great and the good, all with one singular motive in mind, *building the brand image.*

The intended misconception by labels is that when the public see a group on the TV riding high in the charts, the fans automatically assume the members of the group are living an aspirational lifestyle. This financial blessing, however, is rarely ever the case.

Hooky later told me he'd made so little money from Joy Division and New Order, that he had to work as a roadie, loading equipment into the Hacienda just to make ends meet.

Another band who were also starting to breakthrough at the time who subsequently supported both my groups B.F.G. and Lavolta were the Stone Roses. During their heyday in the late '80s the members, I was reliably informed, were paid fifty pounds a week each. It was not the land of milk and honey that we were led to believe.

To add to the New Order bass player's financial problem was the lack of any apparent business acumen by the label boss Tony Wilson. It cost Factory Records more to produce the biggest selling twelve-inch record in history than they received from the distributor. Every single copy of New Order's Blue Monday that sold, Factory Records and therefore the band, lost 5p. This was entirely due to the brains behind the artwork. Peter Saville had designed a sleeve that resembled a computer floppy disk. The cost of manufacturing these sleeves was so expensive that it made the release uneconomical. Tony Wilson, when told about this fundamental business flaw in catalogue number FAC-73, said, 'well nobody's going to buy it anyway'.

If you are still wondering which American band drew around twelve people to a tiny club in Manchester on 25th April 1984, it was REM. I often wonder whatever happened to them.

As time passed, it became apparent Lavolta were in desperate need of a decent manager to make any headway in the industry. The local social secretary at the polytechnic had expressed an interest in the job. We agreed to meet and discuss the matter further. During the resulting discussion, the would-be manager mentioned that there was a ginger-haired kid from a punk band called the Frantic Elevators, that he was also considering managing. The name of the social secretary was Elliot Rashman, and ultimately, he made the right decision. The ginger-haired kid was Mick Hucknall of Simply Red.

That was the beginning of the end for Lavolta. It had run its natural course. My time drumming for the band contained many unforgettable and some unrepeatable memories. Whilst touring with the likes of New Order and the Cult, I never made any money, no one did. I loved every second of it.

It was at this stage that I had to seriously re-assess my life. Things could not go on like this. I sensed if I remained in the bank, the security aspect would slowly suffocate the life out of me. I had to make a move. In December 1985, without any job to go to, I resigned from the bank.

Fortune, as they say, favours the brave, or the stupid and within a couple of weeks, I had my first big break.

I was approached to drum for the band Wax. This particular group formed when the lead singer from 10cc, Graham Gouldman, teamed up with the American singer Andrew Gold.

Wax's biggest hit was 'Building a Bridge to Your Heart'. It was about as far away musically as you could get from what I enjoyed listening to; however, it was an opportunity to do what I loved, and that was to drum.

As well as writing such hits as 'I'm Not In Love' and 'Dreadlock Holiday' with 10cc, Graham had also penned the track 'For Your Love' for the Yardbirds, featuring a young Jimmy Page. Page would later go on to form Led Zeppelin. Graham subsequently told me over a beer that he had been the original choice as bass player for the Blues Rock legends.

The other half of Wax was American singer-songwriter Andrew Gold. Andrew was from musical royalty. His father was awarded an Oscar for writing the soundtrack for the film 'Exodus' starring Paul Newman. Compared to his mother; however, his father was a musical also-ran. Andrew's mother performed under the name of Marni Nixon. Those who have seen the films 'West Side Story', 'The King and I' or 'My Fair Lady', you will have heard Andrew Gold's mother singing. Her performances in those films earned her two Oscars. At one point, Andrew showed me a picture of him shaking hands when he was a teenager with four young chaps who had come to his house to meet his parents. It did not take long to realise who the four suited individuals were. Rumour has it that this particular band went their separate ways because of the Japanese wife of one of the primary songwriters.

Andrew, before he formed Wax, also had an incredibly successful solo singer-songwriter career. He had hits with 'Lonely Boy', 'Never Let Her Slip Away', 'How Can This Be Love,' and the signature tune 'Thank you For Being a Friend' which featured on the hit comedy series the 'Golden Girls'.

Before this success, he had also worked with Linda Ronstadt and the Eagles.

Andrew also had one of the coolest claims to fame possible. I was not even aware of this fact until after he died in 2011. One of Andrew's tracks had been used to 'wake-up' the Mars pathfinder Space probe in 1996; therefore, Andrew could claim to be the first human ever to be 'heard' on another planet.

I could not have possibly known this at the time, but that decision to drum for Wax would change the course of my life in an entirely unexpected way. During the period I was involved, we played numerous television shows throughout Europe, alongside some of the biggest bands of the time. Frankie goes to Hollywood, Depeche Mode, Aha, Simply Red, Elvis Costello and Eurythmics. The most significant appearance we made was at the Montreux Rock Festival. It was televised worldwide and considered the most important music show to appear on.

We arrived in Switzerland via a chartered jet that had taken us from London to Geneva. I shared a bottle of wine with Elvis Costello and his Attractions on the coach from Geneva to Montreux and discussed with Aha's Morton Hackett how a fantastic video had given them their break with, what I considered, one of their weaker tracks. This was all before enjoying the luxury of a suite at the Grand Hotel. Yes, these were heady days. From the Nat West Bank in rainy Droylsden to a view of Lake Geneva from my room in a five-star hotel, all in a matter of weeks, this was more like it.

I still had no money.

Despite the massive television audience of tens of millions, I received the princely sum of £250 for that particular appearance. The truth is, it was such a fantastic opportunity; I would have done it for nothing. What I learned about how the world worked during this period was far more valuable than just a few pounds in the back pocket.

Andrew was hugely experienced within the music business. He was the first person to take the time to explain, in detail, how the industry worked. For that, I will be eternally grateful. He emphasised that to make any money, you had to control the publishing. It was all very well writing and performing the songs, but you also had to control the copyright on those songs. That explained the reason that I made £250 for my appearance on a worldwide syndicated rock show. I had no claim on the copyright on those songs, and that was the rock behind which the big money was hiding.

The next revelation by the official NASA 'waker-upper' was to have an even more significant impact on my life.

Andrew had a phobia. He was terrified of flying, petrified in fact. Over the Alps, during the flight to Montreux, we were hit by lightning. Due to the entire passenger compliment of the BAC 1-11 reading like a who's who of '80s pop stars, the incident instantly became worldwide front-page news. White-faced Andrew said to me 'there is only one person who does more flying than me... And that's the pilots, and I hate it'.

It was after removing the expletives that had surrounded Andrew's comment, that the significance of what he said hit home. It was while taking a shower, getting ready for that evening's TV appearance, everything became crystal clear. I quickly rationalised that if I could accomplish performing at such a prestigious event within weeks of resigning from the bank, it was absolutely within my power to achieve what the school careers advisor considered impossible.

One of the best aspects of drumming for Wax was the number of flights and the amount of travelling involved. If I was getting such a buzz from playing in bands in front of huge TV audiences, I presumed this would pale in comparison with the experience of flying a £250 million aeroplane with over 300 people on board, all of whom had absolute faith in your abilities and decisions. If I was getting such a buzz from hitting a drum with a stick, surely, it would be mind-blowing to take off and land something weighing more than a house. From that very moment, there was no doubt in my mind that ultimately I would become an airline pilot. Yes, I had no money. Yes, I had never flown an aeroplane in my life before. Yes, my A-level results were dire, and yes, I had no idea how I was going to accomplish it. I just knew I would.

Andrew had sown the seed, but it was going to be some time before I could even contemplate embarking on that particular career diversion. I would need money, lots of it. The issue was, I had not become a drummer to become rich. My simplistic view had been if the money came my way doing what I loved, that was a bonus. I loved what I was doing. I was drumming and travelling.

Working in a bank had taught me one valuable lesson. Never again would I do anything 'just for money'. But now, for the first time in my life, I needed a substantial amount.

After my discussions with Andrew, he made it clear that there was plenty of money in the business. Labels, managers and publishers syphoned most of it before it trickled down to the artist. Maybe if I cut out the middlemen?

With the blueprint Andrew had given me for a healthy, wealthy and wise future within the industry, I duly set about forming Attica Records as a protective shell for my own band, B.F.G. This was the only way I could see of ever making, and keeping, the substantial capital needed to fund any future flight training. I certainly was not going to be able to save that kind of money working in a typical nine to five job.

I had come up with a crazy idea, based on the punk ideal, that it was more about attitude than the ability to play an instrument. On that basis, I had formed the fledgeling B.F.G. with my old school friend Keith who, by now, had completed his degree. We decided that as the drums were the only instrument that could be reasonably played by either of us, we would, illogically, buy a drum machine. Having no idea about how to play the guitar, I then set about teaching Keith how to play the bass. To complete the 'we haven't a clue what we are doing ethos' Keith and I asked BJ Williams, a music fanatic with whom I used to work with at the bank, if he would join as the singer. My major influences were Lydon's new bandmate in Public Image Ltd, Keith Levene, and the guitarist from the funk-punk Gang of Four, Andy Gill.

Years later, a well-respected American DJ nailed down the resulting noise that we collectively made. "As intense as Joy Division without the depression" was his verdict.

Following all the advice from Andrew Gold, the first single 'Paris/ Amelia' was released on Attica Records. It sold out within a couple of weeks. Within a month the 12-inch record was swapping hands for £25.00 on the second-hand market. One individual who bought a copy was a young barrister in Canada. His name was George.

Amazingly we were doing something that touched a nerve with people, while, at the same time, getting *on* plenty of others. Perfect.

I had sent a copy of the single to Hooky explaining that since I had left Lavolta, I had formed B.F.G. I asked him, as we were just starting out, if he knew of any groups who needed a support act. My idea was to break us in slowly. Maybe a pub gig in a dingy bar where we could stumble through a few tracks without too many witnesses to our likely unholy racket. I don't think Hooky was on the same page as I was on the matter. Within the month B.F.G. were playing their first proper concert in front of several thousand people, supporting New Order on their sell-out tour.

I was learning the business as we played. Without a manager, apart from myself, through various line-up changes, we went from writing and rehearsing in an attic above Keith's parents' bedroom (Attica Records) to performing in front of 15,000 people at the Loreley Stadium in Germany alongside Bjork and The Pogues. All this achieved in under three years.

Our follow up single 'Western Sky' was the first recording that featured the new line-up. As is always the case when a band starts to become successful, it creates a significant dilemma for those members with full-time jobs. It's a dangerous gamble to give up the security of a career, all for the very faint possibility of lasting success as a recording and touring artist. Both BJ Williams and Keith made the very sensible choice that their careers came first. Although this put me in somewhat of a quandary, there was never any ill-feeling over their decisions. I immediately began searching for replacements. I decided that I needed someone who could play the guitar and interpret my ideas along with a new singer and bass player. Western Sky featured Steve on vocals and 'Baggy' Batheram on guitar and a young Irish lad, Joe on bass. Joe would, many years later go on to appear in the band 'Noize' alongside Shepard Fairey, the graphic artist behind the Obama' Hope' poster and 'André the Giant has a Posse,'

Western Sky managed to attract the attention of a journalist who wrote for one of the most widely-read music papers in the UK, Melody Maker. He had given the track an excellent review in that week's new release section. Seeing an opportunity, I rang him to ask if there was any possibility of, what in effect, would be B.F.G.'s first-ever nationwide interview. He agreed and asked if we had any more material and if so, would we send it down. I duly sent down 'Higher Than Heaven Is', a new track slated to be the next single.

By this point, I was not only writing material. I was also running the record label single-handedly. Add to that I was also managing the band, drumming and, when we recorded, playing additional guitar and bass. I urgently needed to outsource some of the workload. I was, however, not desperate to sign any deal at any cost. If you are desperate, turn around and walk away, it's the first rule of negotiation.

To make the most of my trip from Manchester to London for the interview, I had also arranged to see RCA/BMG, one of the biggest music companies in the world. As I sat in their plush surroundings in Cavendish Square, it was hard not to consider how different the working conditions were for office-based music staff, compared with the average musician. Luxurious big leather seats with Evian water or designer coffee on call for those working at a major music company. Compare this to a Pot Noodle and Vimto for the majority of musicians at the coalface, the ones who generate the music that the industry chews up, profits from and then, ultimately, spits out.

I had waited at least 20 minutes in reception at RCA for the meeting to commence. Eventually, I was invited up to the hallowed offices of the A&R (Artist and Repertoire) Department. From the cassette of 'Higher than Heaven' leaving my hand, being placed into the tape machine, play button being hit, stop button being pressed, and the recording being back in my grasp, took all of twenty seconds. The female A&R scout spoke. 'It's not really for us at this stage but if you have any more tracks, feel free to send them to us.'

Translated into A&R speak, this means, 'I could have had another ten minutes lunch break, now be a good chap and bugger off'.

Thirty minutes later and I was sat in a pub next to the Melody Maker offices with a journalist by the name of Mick Mercer. Any disappointments from my meeting with RCA swiftly evaporated. Mick informed me that not only did he love the new track that I had sent him, but he thought it was the best thing he had heard all year.

As I sat on the National Express coach on the way back to Manchester mulling over the day's events, I could only hope what Mick had told me would somehow find its way into print if, and when, the interview was published. After all, I had not completed the usual label - journalist transaction of buying him a four-course lunch, wine included, to ensure the inclusion of the interview in any forthcoming issue. I need not have worried.

"This Year's Greatest Independent Single is B.F.G.'s Higher Than Heaven" was the headline when the article surfaced a couple of weeks later. The reputation of the music business is that of a corrupt nest of vipers. In a few instances, this is very true; however, most are in it for the love of the form. Mick Mercer was one of them.

The afternoon that the interview came out, the phone rang. It was the same person who had given me less than thirty seconds of her time in Cavendish Square. The conversation took place many years ago, but it will be forever etched in my memory -

'Hi Mike I have not been able to get that track out of my head that you played me, we would love to come and see B.F.G. play live. When are you next playing?'

I felt no need to be accommodating;

'We don't have any gigs planned.'

'Well, can you not arrange one in London as we would like to come and see you?' A typical opening gambit by a label interested in an artist.

My response was not what the majority of artists would have given.

'Nope, I'm not doing that as it's a lot of effort to go to for you not to show up.'

She promised that she would but then said, 'Ok, if you don't want to organise a gig, we will come and see you rehearse'. Then she proceeded to recommend some good rehearsal rooms down Holloway Road in London.

I was confident that they were interested and on the hook, but at the same time, I had not forgotten the arrogant treatment afforded to me when I visited the RCA office. The power in this exchange was now mine. 'Holloway Road is in London. We are a Manchester band.'

'I don't think you understand we are one of the biggest companies and we are telling you we want to come and see you.' The RCA representative was now getting a little irked. Maybe she had failed to understand the changing dynamics of the situation. She could not have made it any more obvious that she was keen to sign us if she had waved a cheque in front of my face. She was desperate to get to the band before any other company came in with an offer that I could not refuse. Rule number two in negotiation, never appear grateful or desperate.

'We rehearse in Manchester.'

'Right,' was her terse response. This conversation was not going to plan at all, at least from RCA's point of view. It was a game of poker.

There was a pause. Rule number three, never talk during a negotiation just to fill in the silence. She now knew I was not desperate. The swanky London address, receptionist, leather chairs and Evian had not done the usual trick of convincing a penniless musician that they needed RCA more than RCA needed them.

'Err. Mike! Still there?'

'Still here.'

'Ok, ok, what day do you rehearse?'

What the representative of one of the most successful music corporations in history had not been made aware of was, at this stage, B.F.G. was now not a functioning live band. Joe, some weeks previously had decided to return to his native Northern Ireland. I was now therefore into the realms of mental gymnastics as to who we could ask to play bass and how many tracks they could learn before I could offer a date for RCA.

A name suddenly came to mind. 'Week Wednesday.'

With that, a date with the devil was struck.

Unfortunately, the standby bassist was not available until the day before the RCA entourage arrival. Still, I raced through as many of the tracks as I could with him.

The day of the RCA visit eventually arrived, and the high-powered congregation finally pulled up outside the rehearsal room in Chorlton cum Hardy we shared with Simply Red. As soon as the kingmakers took their seats, we blasted through fifteen minutes of music then stopped dead.

I announced from behind the drum kit; 'Thanks for coming. I hope it's been useful for you'.

The assembled music execs looked at each other nonplussed. They were less than impressed and made it clear it was unlikely I would be placed on their collective Christmas card list any time soon.

'What do you mean? We have come all this way to see you rehearse. You do have more songs, don't you?'

We did have more tracks, but the bass player had only managed to learn four or five well enough to play to an audience. It was time to play them at their own game.

'I am sure you are aware there was an article about us in Melody Maker. It's caused a lot of interest so we can only give you fifteen minutes. We have more companies coming to see us tonight.'

After a few furtive looks and whispered comments between the assembled A&R managers, the head honcho finally enquired, 'who else is coming?'

'Do you want me to tell them you were here?' was my response.

At that, the woman who only a couple of weeks previously had kept me waiting for twenty minutes turned to her fellow travellers from the capital, 'well, it was worth a try.'

Dumbfounded, they stood up in unison and made their way to the door that they had walked through only fifteen minutes earlier. 'We will be in touch.' was the departing line of the disgruntled audience.

Immediately they drove away there was consternation in the B.F.G. camp.

'You've blown it!'

'What?'

'We had a chance there, and you've screwed it. Do you think they are going to sign us after that?' The band were not impressed. I, however, was delighted. By the close of play the following day, a deal was offered, leaving us with total artistic freedom and a sizeable advance. The important aspect was that I was now able to outsource some of my workload. This allowed me to concentrate more on the reason I was in the industry in the first place, to make music. To look after the financial aspect, the first cheque I wrote when the advance hit the Attica bank account was issued to a leading Manchester accountant.

Over the following few months, I got to know the lady who I had dealt with at RCA pretty well. In reality, she was a charming individual. We both knew this was just how the business worked. If you did not realise your value, then you had no place in the music business. It would eat you alive. That's why the vast majority of artists never see any real money. She later told me that during a meeting with her CEO, they had discussed how the negotiations had gone. One of the main reasons that RCA signed B.F.G. was because we appeared to know what we were doing. We were a safe bet.

At this point, you may be wondering why I am going into such detail about my previous life within the music business. The reason – despite the outward appearance of little similarity between the music business and the world of aviation, in reality, the two are very similar. Both feed off the general naïveté of those wanting to become involved in either industry. Both are in the business of selling dreams.

An example, at least from the music business prospective was, and still is to some extent, based on the desperation of the vast majority of people trying to break into the industry. Most assume if they land a major record deal, then they have 'made it'. Most who sign a major deal have no idea of the financial hole which they have effectively now just placed themselves in. The record deals offered in the '80s and '90s make the current payday lenders APRs of 2000% seem like the deal of the century.

A record label, when it signs a band, pays that band an advance. What most fail to realise is that this is nothing more than a loan. From that loan, first, the taxman wants his cut. The manager will then take around 20% gross of the advance, thereby taking advantage of the fact that a majority of musicians are not aware of the difference between gross and net. Of course, it is not in the manager's or label's interests to explain the difference. From what is left, which is now less than 50% of the original advance (loan) the band have to pay for equipment, rent, food and in some cases studio time. Sometimes studio costs are fronted by the label, but, that amount is added to the loan, as are the expense of videos and tour support.

These facts are all in the contract which contains a good sprinkling of the following: 'cross-collateralisation' and 'recoupable'. Mention those words to the typical musician, and it will usually elicit a blank stare and an open mouth. Those two words, however, are fundamental to how the music industry operates. Recoupable means that all the money spent by the label on a particular band is added to the original advance to be paid back through record sales. Cross-collateralization ensures that the label can access money from any of the artist's operations such as touring, merchandise, or a very successful release. They then 'recoup' the entire advance paid to the artist for the less successful releases. The band will generally repay the advance from sales.

So, how much of a financial hole are most artists in?

The following is just a quick peek behind the curtain that Andrew Gold pulled back to reveal the reality of being a musician. There are, of course, a tiny minority who make millions, but you would have more luck playing the lottery. The figures make for grim reading. Assuming the band get an advance of £250,000. Of this, the manager takes his 20% gross £50,000.

The loan (advance) must be repaid out of the royalty rate agreed in the deal. Typically 13% of 90% of retail. At this point, you will have noticed the band only get royalties on 90% of the albums they sell. That is because the label deducts 10% for the cost of packaging. Let's assume that the band's first release is a successful one, and it sells 1/4 million copies. The average cost of a CD in the shops is, in round figures, £10.00. The amount of money generated by those sales is therefore around 2.5 million pounds. For a band of four individuals who have agreed to split income equally, how much would you guess each band member would earn at the end of that year? £200,000, £300,000, more?

Welcome to the Alice in Wonderland world of the music industry.

13% of 90 % of sales = £292,500. With the advance of £250,000 repaid that leaves a profit of £42,000. Well actually, no it doesn't. There are legal fees, studio time, musical equipment, producer's fees, equipment rental, mastering, video costs, stage wear, promotion.

First and foremost, the government will want its pound of flesh in the form of tax. The list of outgoings goes on and on whereas the list of income streams shrinks year on year.

When the accountants at year's end come to audit, the final figures for each 'cog' in the production of the album, after the distributor and retail have taken the lion's share of the £2.5 million will be in the region of:

Record company: £700,000 Producer: £100,000 Manager: £50,000

Studio: £100,000
Booking Agent: £10,000
Solicitor £20,000

As you can now see, the band, at the end of the year where they have made the 'music industry' over £2.5 million will each be lucky to earn half that which a BMI cabin crew receives. It's worse now than how I have described it. For example, up until recently, major labels were holding back money to be paid as royalties due to breakages on downloads and streams!

As if that was not bad enough, in the above example, the band will, after paying all the costs, still actually owe money to the label. i.e. They will not have recouped. The follow-up album, driven by the success of the first, is promoted even more heavily. Again, this is ultimately paid for by the band. If any of the band's albums become massive sellers, the group will be the last to see any money due to cross-collateralisation.

If you're still not convinced, ask yourself why, after selling millions of albums and selling out stadiums worldwide did Michael Jackson die broke? He was not the first and will not be the last to do so. What many artists fail to realise is, despite eventually paying for everything, they never own what they made. The label does.

Remember Take That? Well, take this. The group were on half royalties whenever their label spent more than £25,000 on promotion in any month. No guesses when that particular promotional budget was spent?

Wham's first label deal they signed gave an excellent royalty rate on singles, but the two-piece group would get zero for 12-inch releases. The guys assumed that as the number of 12-inch single remixes were not going to be big sellers for them, they would agree to the deal. Unfortunately, they had not seen the catch. The only real money in the 80/90's was in album sales. And what size are albums? The list of bands not doing 'due diligence' on deals is endless. It's still going on. Spotify anyone?

While researching for an artist on the newly re-launched Attica Records, I was reviewing the latest way that people listen to music. Spotify is currently the only show in town.

The artist I am currently working with is the pianist Amelia Chain. The music is classical in style, piano and orchestral, similar to the music of Yann Tiersen, and Ludovico Einaudi. As I write this, I am sat next to her on the tour bus.

During research for the Amelia Chain release, I found that Spotify pays, in very approximate figures, $0.004 to the copyright holder for every stream. Now consider that if you are on a major deal, with a royalty rate of 13% of 90%, this equates to $0.000468 per stream. For every million streams the artist/group is paid $468. The label pockets $4000. It's going to be a long time before £250,000 is recouped.

The only way for an artist to gain a significant number of streams on Spotify is to have a track appear on a Spotify-curated playlist. Each of these playlists can have over one million monthly listeners. As I researched these playlists, I was coming across artists with tens of millions of streams that not only had I never heard of, but they had no social media presence whatsoever or even a website. All very strange. They all appeared to live in an ecosystem that was Spotify and Spotify alone. Shortly afterwards a story broke in the music industry magazines that accused Spotify of having fake artists on the service. It was alleged the company was paying a flat fee to session artists. They would then place the tracks on one of their playlists, ensuring regular income for the platform with no further outgoings to those who had written the music. Spotify, as you would expect, denied it.

The players might be different, but the game never changes.

It is estimated that 95% of bands that have signed to a record label never see any more money other than their very first advance.

This is now, but back in the eighties, we still had no real money to show for our efforts. Everything I made was being reinvested in paying for studio time, new releases, artwork, promotion, and covering the costs of the large number of concerts we were now performing abroad. No money yet, but my eye was on the potential, I was building it slowly and carefully.

The Stone Roses were beginning to break into the big time along with a host of other Manchester bands such as The Inspiral Carpets, 808 State, James and Happy Mondays. B.F.G. was tipped to be the next band in the *Madchester* explosion. Everything was going to plan. Everything was running like clockwork. Everything was going so swimmingly...

The letter from the receiver explained everything. The distribution company that I had signed Attica Records to had gone into liquidation. It owed a lot of money to a lot of bands and record labels. Unfortunately, Attica was one of them. Without the money owed, it would be impossible to carry on. All those record sales and concerts, would, in the end, count for nought.

A meeting was called for all of the creditors to attend. I managed to find myself sitting next to members of the band The Wedding Present. They looked as miserable as I felt. All that work, time and effort, all for nothing,—of course, we had had a fantastic amount of fun, but nevertheless, it left a sour taste in the mouth. The affairs of the distribution company that handled our releases were inextricably linked to the problems experienced by the record division at Rough Trade. Rough Trade was the record label for whom The Smiths recorded. Rough Trade owed The Smiths a significant amount of money. As the label had folded, The Smiths had now signed EMI. Their new label had the legal muscle to squeeze every last penny owed to their new protégés. That would mean both myself and The Wedding Present, among many others, would be lucky to walk away with anything more than a soggy packet of crisps. Even then, we would probably have had to share the packet.

It looked like it was back to square one, no money, no band, no car, no house and no label. I did, however, gain one invaluable ability. It was the ability to sense bovine excrement at one thousand paces. Several years of listening to coke-addled individuals with bubble perms and nasty silk bomber tour jackets, usually with the sleeves pushed up to the elbows had honed this skill. They would promise the earth but would deliver nothing. Years of hearing every excuse under the sun from promoters and labels and distributors as to why they couldn't pay money owed only served to heighten that sense. Of course, this tactic of business refusing to honour written contracts keeps the legal profession in gainful employment and not, as I was to learn years later, just in the music business. It still amazes me how often this tactic works. It appeared that in the music business, as it did in later life in the aviation world, individuals just seemed to shrug their shoulders and give up chasing what was rightfully theirs.

Twenty-eight years old, without a penny to my name. Most of my old school friends were now doctors, dentists, solicitors, and most had children and houses with cloakrooms bigger than the rented flat I was still occupying in Didsbury.

I had grown tired of relying on people who frequently did not uphold their side of a bargain, regularly letting me down. I was continually questioning peoples' motives and integrity. But this *was* the music business after all.

The financial collapse of Rough Trade had not only shattered my future in music, but it had also smashed into tiny pieces any long-held dream to finance a future career as a pilot.

I was still shell-shocked about the turn of events when I decided to take a bus ride into the city centre. This was a few years before that well-known demolition company, the provisional IRA, thankfully decided that Mancunians were sick and tired of the Arndale Shopping Centre. The Arndale was a retail complex in the heart of Manchester that visually resembled a public toilet. It also shared the same aroma. The building was as bland as a service station sandwich and was the same shade of brown that resembled the meat filling within.

The Arndale's one saving grace was that it had a branch of the Waterstones bookstore located within its sprawling reaches. Being a typical Manchester summer's day, it was cold, and it was wet. It was pouring down. To take shelter, I dashed into the bookshop along with what felt like every other person within a fifty-mile radius. The resulting musty smell of a group of soaked shoppers resembled that of a damp dog. It was a slight improvement on the previous Arndale fragrance. As I stood there waiting for the downpour to ease off, out of the corner of my eye, I saw a book that this time really did light the blue touch paper. It was a flame that had threatened to catch several times in the past but had never progressed past smouldering. The title of the book escapes me, but it was one of a series aimed at preparing pilots for their ground school exams. This one, in particular, was concerned with meteorology. The memories of what I wanted to do as a child, my dismissive careers adviser, and the inspirational conversation with Andrew Gold all came flooding back to the forefront of my mind, collapsing in a heap in my frontal lobe. I picked up the book. My life would never be the same again.

# CHAPTER THREE

## *Waterstones Watershed*

*"No more snarly up-wordings."*

Professor Stanley Unwin (Amstrad advert)

I had already flicked through several pages of text and diagrams in the pilot study book whilst still being jostled by damp shoppers. Eventually, I came to a page that contained a revision test. It just so happened that the questions were about onshore and offshore sea breezes. I remembered many years previously, my father, a keen dinghy sailor, explaining the cause of this phenomena. Let's see if I could recall what he had said to me in the dim and distant past. Ten minutes later and I had achieved eighteen out of twenty in a pilot exam. Maybe I was not as dim as my exam results as a teenager had led me to believe. Perhaps this was just a flash in the pan, beginners luck.

Did success in an exam on sea breezes in a very damp Waterstones indicate that my childhood dream of flying aeroplanes had been within my grasp all along? Did that 90% mean I could compete with those heroes of the skies, trained at Her Majesty's expense with whom I would be competing with for a chance to sit in the pilot's seat of an airliner?

Come on, who was I trying to kid?

All airline pilots were intelligent, super fit, and usually had a double-barrelled name, all traits which were sadly lacking from my canon. The one thing that I was sure of was that all pilots had an unwavering self-confidence to *do the right thing*. After all, they had the responsibility of hundreds of peoples' lives in their hands, not least, their own.

Buoyed by my stellar exam result, once the rain stopped falling for a few minutes, I dashed, for the local WH Smiths to leaf through any aviation magazines that they might stock. The first one I grabbed was entitled 'Pilot', a title that would minimise any confusion as to what subject matter was within its pages.

As I flicked through its contents, I came across an advert for a Commercial Pilot's License (CPL) course, without an Instrument Rating, all for the total of £36,000 at Coventry airport. It was certainly cheaper than I imagined, but it was still way out of reach for a drummer on the dole. If I was serious about becoming a pilot, how was I going to get my hands on £36,000? The Instrument Rating, required for any serious commercial flying, would be at least another £20,000 to £30,000. To give some idea of the relative costs, at the time, a decent-sized flat in central London would cost around £50,000.

The next day, for the first time in my life, I walked through the door of Manchester's Central library intending to make the most of my visit to this pre-Google depository of knowledge. I searched for any information on how to advance from bank clerk to a drummer to then operating one of humanity's most outstanding engineering achievements, all on a shoestring.

Several ways were suggested in the careers pamphlet. The first was to join the RAF. This particular avenue I had discounted years ago, based on my view, I would be 30 when I eventually left to pursue an airline job. But here I was, only two years short of that age with zero flying experience and zero income.

The second method was somehow to gain a scholarship with an established airline. At the end of the course, assuming the airline still needed pilots, there would be a position waiting. The dilemma with that particular avenue was that most of the major airlines at that time were not offering scholarships, and those that did were hugely oversubscribed. Virtually every applicant would have a degree. Most would also be significantly younger than me. For those lucky enough to be accepted, the cadet still had to secure the funds to pay half of the training fees upfront with the airline then advancing the rest in the form of a loan to the cadet. The cadet would then be required, for five years after completing the training, to pay the company loan back. The money would be taken directly out of their salary. That sounded exactly like the kind of deal a major record label would offer.

The final method was the self-improver route. It was significantly more complicated than a scholarship. Still, it had the advantage that once you achieved a certain level, you could start earning money as a flying instructor before you began to work on an aircraft that you could walk underneath. It worked like this:

The first stage was to gain a PPL, also known as a Private Pilot's License. The number of hours required to pass this stage varied. The minimum was forty hours. The average for someone only flying once a week was around sixty. With your shiny new PPL, you then built your flight time to over 130 hours to allow you to sit the commercial pilot written exams of which there was around fifteen or so, all on different aspects of the profession: Navigation, Meteorology, Air law, Instruments etc. You would then complete an IMC (Instrument Metrological Conditions) course, which was a minimum of fifteen hours. IMC was a necessary instrument course designed to get you out of trouble if you inadvertently entered clouds while flying. Next up was a Basic Commercial Pilot's flight course of around fifteen hours and then a flight test. Finally, before you could earn a penny, was a 28-hour Assistant Flying Instructor course.

Even after all that flying, at the cost of anywhere between £70 to £100 an hour, you still had a very long way to go before you would be able to apply to the airlines for a first officer position.

The good news was, once you gained your flying instructor license, you could earn around £8.00 an hour. Yes, it was a pittance, but at least you did not have to pay the £70-100 per hour to build your total flight hours to seven hundred. It was at seven hundred hours you could be issued a Commercial Pilots License.

If you think that was it, think again.

At that point, there are another twenty or so written exams for the ATPL (Airline Transport Pilots License). The pass mark was 75% and negative marking if you got a question wrong. Fail more than three of the exams, and you are re-sitting the whole lot. Add to that, before an ATPL can be issued there is the requirement for at least 1500 flying hours.

Even then, you are still not finished.

There was still one final sting in the tail, without which there would be no flying a sleek shiny jet, or indeed a battered turboprop delivering the night mail.

To do any of the above required an Instrument Rating (IR). The training would cost around £5.00 a minute. The high cost was because the IR training and test had to be flown on a twin-engined aeroplane, which came with significantly higher running costs than the previously used single-engine aircraft.

Before I commenced this journey, I needed to know if I had the necessary academic ability required. The two failed A levels did not present a solid case that I did.

I paid a second visit to the library and withdrew two books, both aimed at those who were about to embark on the study for the highest level of pilot's license the ATPL. It was the equivalent of taking a book out about quantum physics by somebody who had only just learned to read. However, my thinking was sound. If I could get my head around the contents of this book, all I then needed to worry about was the little matter of how to pay for the training.

Although ambitious, I was entirely realistic in my expectations due to my late start. There would be no way on earth I could ever conceive a first officer's position on a passenger jet. Those jobs were reserved for the ace of bases with an aeronautical degree. Flying a turboprop delivering night mail would be my Concorde or 747.

I raced down the steps of the library in my torn jeans and leather jacket eager to get back home and make a start with the two ground studies for pilots books safely tucked underneath my arm. On the other side of the road, coming out of one of Manchester's top hotels were two uniformed pilots, they were surrounded by five immaculately dressed women, also in uniform. The logo on the side of their transportation gave away the name that would appear at the top of their undoubtedly substantial salary cheque. It was a British Airways crew who had night-stopped in Manchester and were now obviously flying off to some exotic destination, probably London Gatwick. In those days even Gatwick sounded like a far-flung destination.

I virtually leapt up the stairs to my third-floor flat so I could get my head into the first of the two books that I had relieved from the Manchester library. The book was by RD Underdown. The topic was "Navigation". I was going to find out within the next few hours if someone who had spectacularly flunked his maths A-level would be able to deal with the academic requirement to make head or tails of this particular and technically challenging study book. It was either going to be a spectacular car crash or a revelation.

I opened the first page in a thick red notebook I had bought specifically for this enterprise and wrote the headline NAVIGATION and underlined it. As I opened "Ground Studies for Pilots" I noticed that the book had only been checked out of the library on two previous occasions. Maybe the competition to become a pilot was not as fierce as I had first feared.

The first section was concerned with maps and charts. Very soon, I discovered that Great Britain was not the same size as India, although that appeared to be the case when viewed on a map of the world. This was due to scale distortion caused by the projection of choice of the British Empire. It was known as a Mercator projection. Maps constructed in this manner distort the size of landmasses and oceans as the latitude increases from the Equator to the poles. It delighted those wishing to give the impression to the rest of the world that the small wet island off the coast of France was far more imposing than it actually was.

I had scratched my head in my maths' A' level class, stumped and considered it a very far-fetched idea that I would ever need to use the cosine or sine rule at any point in my life. Yet within thirty minutes of sitting down with my booty from the library, here I was, working out the track and heading required to get to a particular point on the surface of the earth. No teacher, no distractions, no pressure, just working from first principles. Importantly, I was getting the questions right. Far more importantly than that, I completely understood the principle behind what was being examined. I felt like a sponge that had been left out to dry for far too long. Knowledge was just pouring into me.

Along with the seemingly never-ending technical books, I also got my hands on a volume so large it doubled as a windbreak. This book was the entire private pilot's license syllabus written as a guide to help flying instructors teach and get across the basics of how to fly an aeroplane to the ugly ducklings. All I needed now was an aeroplane to practice on and put the theory into good effect.

Although I did not have an aeroplane, I did have a computer, an Amstrad 9512, which boasted a full 512K of memory. To this day, this computer is the only one that I've ever had that never crashed.

It was in the back of 'Pilot' that I saw a flight simulator software package for £50. I calculated if I survived on baked potatoes for a month, I could afford it. I lost about five kilos but gained a floppy disk containing the 1's and 0's that collectively powered one of the first computerised flight simulators sold to the general public. The frame rate was so slow I could have sketched the images quicker than the tiny Amstrad graphics card could re-draw it. Ultimately, however, that £50 piece of computer code saved me tens of thousands of pounds in training costs.

The hours sitting in front of the Amstrad allowed me to practice what all pilots need to become proficient at - their scanning. The necessary aircraft instruments are arranged in a T-shape, the dominant instrument being the attitude indicator, through which all scans must pass. Attitude, glance to the left, check the airspeed, back to the attitude, check the wings are still level, down to the heading, back to the attitude, over to the right to the altimeter, make sure we are still holding the correct altitude. It's the aeronautical version of spinning plates.

A month went past, and it was again time to exchange the books at Manchester's central library. Once again, it was raining heavily. The bus back from Manchester was full of damp students, providing a canvas for children of all ages to attempt to depict, as best they could, anatomically dubious outlines of the male reproductive organ. All etched with a solo digit in the condensation that had formed on the windows.

I elected to be content with just being able to see out of the window. It was a challenge. The bus slowly made its way up Oxford Road and past where my story began, Manchester St. Mary's Hospital, where I had been born twenty-eight years previously. As we passed the building on the left that I had said hello to the world from, on the right, a branch of Barclays bank came into view with a large poster in its window advertising 'Career Development Loans'. Even with the restricted visibility through the bus window, the poster screamed 'opportunity'.

The next moment I was stood in the rain watching the artwork laden transport disappearing into the mist. The last part of the puzzle was now starting to take shape.

Walking into the branch, the smell of money took me back to my job at the Nat West Bank.

A middle-aged gent introduced himself as the career development loan officer. He explained the loan mechanism and that the maximum amount was £5000. Although this was just a fraction of what I needed for the entire project, it would allow me to at least secure the Private Pilot's License, all the written exams and, if everything went well, the commercial flight training. A date was set for a formal interview to determine if my dream was considered suitable for a loan.

I virtually ran the four miles home. I barely noticed the lashing rain as I considered how to proceed tactically. I would have to ditch the leather biker jacket and source a pair of trousers that did not expose my knees to the world.

Given my age, my lack of a degree or even decent A-level results, my only option was the self-improver route. I had to be able to complete the training as inexpensively and as swiftly as possible to be in the position to begin to pay the loan off as I built my hours as a flying instructor. Somehow I needed to be able to convince the bank that the £5000 would cover everything up to and including the flight instructor course.

I scoured the back pages of 'Pilot' magazine for the absolute cheapest aircraft hire. I was prepared to travel anywhere in the world if the price was right. I failed to factor in unnecessary expenses, such as the cost of the airfare to get to the most affordable aircraft hire, or any money that I would need for such irrelevant necessities, such as accommodation, food and water. I could always eat grass, drink rainwater and live in a tent.

I was working on the principle that the suit who would decide if I did or I did not receive the loan, would be on a commission. If he did authorise this financial lifeline, it would probably mean a nice bonus in that month's salary packet. Losing some of the costs would, I am sure, win me plaudits from the higher echelons of the Enron and Thomas Cook accounting departments but it made it easier for the man from Barclays to say yes.

It was even easier than I could have ever dreamed. I walked out of the Barclays Altringham branch a couple of days later with the promise that £5000 would be in my account within the next three days. The omens were excellent, and the sun was shining for once.

Even before I had the meeting with the bank, I had paid a visit to the local flying school that was based on a grass airstrip several miles north of Manchester airport. I needed to know what the baseline cost for the PPL would be and how long it would take. Forty was the minimum number of flying hours you would need to log before being issued with the license.

The CFI (Chief Flying Instructor) seemed to enjoy, just a little too much, informing me that there would be no way of completing the course in forty hours. He blamed the weather.

It was time for me to retreat. 'You make an excellent point, so I think we'll leave it there.' The look on his face made it clear that he understood he had just talked himself out of gaining another student and several thousand pounds.

I cycled back home, complete with rubber bands around my ankles to prevent my jeans from becoming ensnared in the chain and becoming even more ripped. Yes, financially things were as tight as the rubber bands.

It was a day or so later, flicking through 'Pilot' magazine that I noticed an advert in the classified section for the "Squadron Two" flying club based in San Jose, California. The advert looked impressive and the cost per hour seemed too good to be true, and in life, if something seemed too good to be true, it usually is a dud.

It was worth a long-distance phone call at least.

'Hi, Frank here.'

I was surprised to find myself talking to someone with an English accent.

I told him exactly what I said to the defeatist CFI.

'It's all up to you. If you put the work in, of course you can finish and pass in forty hours. Plus, the weather here is virtually always CAVOK.' With my new-found knowledge, I was now able to trade abbreviations with the best of them.

'So mostly VFR days, and not too much IMC?'

Aircraft hire in San Jose was less than a third of what it would cost in the UK. The nagging thought in the back of my mind was 'there must be a catch'. I rang a few of the numbers that Frank had provided as recommendations. Everyone remarked how friendly and thorough the training was. The only problem appeared to be that the FAA would issue an American pilot's license upon completion of the course.

In 1989 I asked the CAA an exceedingly straightforward question. Can I exchange an FAA PPL for a UK PPL? I had an immediate straight forward answer from the CAA. 'Yes.'

All I needed to do was pass the English air law exam.

San Jose was going to save me an absolute fortune. Things were falling into place.

The air law was passed with no issue. There was now just one last major obstacle that could derail this dream before it even took off. It involved someone cupping the family jewels and asking me to cough. It was the CAA Class One medical.

I had always assumed, as most do, that airline pilots had to be super fit with eyesight like a hawk and be able to hear a wallet being opened from the other side of the bar. I would soon find out.

I booked the medical at the CAA's main centre at Gatwick airport, then spent the two weeks cycling every day between my flat and

Manchester Airport. What I thought I could achieve in two weeks cycling around ten miles a day I am not sure, but it got the heart pumping. At least I was making an effort.

Two weeks later, I duly arrived at the CAA medical centre with a certain amount of trepidation and promptly was relieved of yet another £200. Bumps felt, eyes thoroughly tested, hearing checked and a host of other examinations resulted in a happy ex-drummer walking out of the CAA reception area. I now was the proud owner of a white piece of paper that announced I was medically fit enough to command a Boeing 747. I just needed to learn how to fly one!

Now that I had the medical under my belt, I just had one thing to complete before I jetted off to begin the next chapter of my life, and that was to complete the previous chapter.

I had booked my flight to Oakland in California and would be leaving in a month. This would give me enough time to write and record another B.F.G. album!

Shifting my focus from dreams of flying, back to music, I wrote the entire album in two weeks. Lyrical ideas were generated by grabbing random books and turning to random pages and taking random sentences and then taking them totally out of context. Working with my old friend and original B.F.G. vocalist B.J. Williams we were ready to record the album by the end of the fortnight. One advantage of your vocalist still working at a bank was that he could approve a business loan to cover the recordings and manufacturing of said album.

I had decided to use the same studio in Leeds where I had recorded the previous LP, Fathoms. Lion studios was owned by two musical geniuses, Guy and Andy. I considered it a fitting end to my musical career to work with these two gifted individuals again. Recording in Leeds had the added advantage that I could use the train journey from Manchester to revise for my American written pilot exams. This particular exam was in the form of multiple-choice questions taken from a big thick red book. The task was to know the book inside out, practice the questions, and you would pass the FAA PPL written exam.

For the first week in the studio, I took the musical sketches that I had made at home and made them into bass lines and guitar parts that were then layered over thumping drum machine kicks and snares. To make things interesting, Guy suggested embracing sampling technology which at the time was in its infancy. We reserved the second week for trying to chisel out from this musical block of granite, something that would be recognisable as a melody. Halfway through the last week, BJ Williams arrived for one day to complete the vocals for the entire album. We had to seriously consider cancelling the session when, a few days before he was due to record, BJ announced he was feeling rather ill, but, in the true spirit of rock and roll, he said he would soldier on. By halfway through the recording session, it became apparent what the vocalist's malaise was. He was either going through another bout of puberty or, more likely suffering from chickenpox. While BJ scratched, itched and squirmed in the vocal booth, those of us in the control room discussed if any of us had ever had chickenpox.

We concluded that, while we worked out our respective childhood medical history, the vocal booth would serve as an excellent isolation room. We also reasoned that considering BJ was currently in the vocal booth, he might as well carry on with his vocal performances. His vocal takes were astonishing. Maybe it was because he knew the heartless producer would not let him out until we had the perfect take.

The resulting album was BLUE. It was one of the very first releases that would come to be known as 'Industrial' music. Groups such as Nine Inch Nails and Ministry would later explore this genre of music with significantly more commercial success a few years later, but at least we had planted the flag.

A full thirty years would pass before that album would be played live, but more of that later.

I had concluded that I had previously had chickenpox in my teens, so I was safe. San Jose, here we come!

# CHAPTER FOUR

## *I Know the Way To San Jose*

*"If in doubt, there is no doubt."*

Anonymous.

San Jose was even better than I could have ever dreamed, apart from one thing. What I had initially put down to my first experience of jet lag was actually, courtesy of a contagious vocalist, the announcement to the rest of my body that I had contracted chickenpox. The blisters swiftly followed. My first week in California was spent itching. The enforced delay to my new career did, though, allow me to rehearse every lesson in my head. The thought of finally being able to get my hands on a real aeroplane was the best incentive I could have for getting better. I had to get better, quick. I did not have the funds not to.

Even now, in sunny California, I still could not entirely escape my musical past. The very first day I had the all-clear to venture out and mix with the locals I went to cash my traveller's cheques at the local bank. I came out, not only with a bundle of dollars but also the cashier's phone number. The young lady in question had remarked on the T-shirt I was wearing, which promoted a festival where B.F.G. had performed. She had commented that she was a fan of some of the bands that appeared.

It turned out Meg was precisely the right lady to meet at precisely the right time.

That meeting led to a wild three weeks of getting to grips with a Cessna 152 during the day, and, thanks to Meg, seeing bands and partying in the evening. At one such party, I chatted to a chap who was a bit older than the rest. We talked about flying and music for a good hour or so. Afterwards, Meg enquired did I know who I had just spent the time talking to. I had no idea that it was the 'Legendary Star Dust Cowboy', the individual who had inspired David Bowie. He was the man who Bowie named his alter ego Ziggy Stardust after. The following night I was sharing a beer with a Dead Kennedy. I hasten to add this was not a séance; it was with the bass player from the legendary punk band of that name. He was yet another friend of Meg.

As to the flying, it was magical. My flying instructor was to be Julie, the daughter of Frank, who owned Squadron Two. As I waited for Julie to arrive, I noticed a stunning blonde lady with very high cut Daisy Duke-style jean shorts on. She was also wearing a loose-fitting T-Shirt that completed the laid back vibe. 'Hi, are you Mike?' It was clear Julie had not dressed for the beach, but the first flying lesson for the would-be English pilot. It was going to be a hellish few weeks, said no one.

Despite the glamorous instructor, the first lesson was horrendous. From taking off from the Northerly runway at San Jose until we landed, we were thrown around like a penny in a tumble dryer. My knuckles were white as I held onto the seat as if my life depended on it. Maybe it did! The little Lycoming engine tone changed as the huge up and downdrafts that were throwing the Cessna around changed the relative angle of attack of the blade of the propeller to the air it was currently slicing through. I glanced over at Julie.

'Enjoying it?' she asked, laughing.

I nodded back as if this was an everyday occurrence. I may have even forced a grin. It probably looked more like the face of someone suffering from haemorrhoids.

'It's always like this at this time of the day.' Bang! We hit another shear of rising air. It may as well have been concrete.

Moments later, I was let loose. 'You have control.'

'I have control.' At last, at twenty-eight, I had finally taken control of a real aircraft. It felt more like the real aircraft had control of me.

It was shortly after the ninth hour of instruction and bouncing around flying touch and goes on runway 30R at San Jose that Julie asked me to taxi back to the apron. It had only seemed like thirty minutes had passed since we started the lesson, but Julie was the one with the watch. As we approached the turn for the parking area, Julie announced 'Pull over here and keep the engine running.' After unbuckling her seatbelt and before she jumped out she tapped me on my arm 'One circuit OK?' It transpired it *had* only been thirty minutes.

Yep, this was it, the moment I was going to be flying solo for the first time in my life. I had no mental capacity left to feel nervous. Every single synapse was concentrating on the task in hand. Truth be told I can't remember much about that first solo apart from after I had concluded my 'downwind' memory item checklist, I burst into singing the chorus to Higher Than Heaven.

Despite the distraction of having a flying instructor who was as laid back as her attire indicated, as her father had predicted, I passed my Private Pilot's flight test after forty hours of flying. The train journey between Manchester and Leeds had also not been wasted. I managed to pass the FAA ground school exam at the first time of asking.

What followed was six weeks of flying the length and breadth of California, sleeping under the wing of N25562 while I built my hours. I was finally living the dream. There is an entire book to be written about those exploits on their own at some point.

The airfields I was flying into could not have been more varied. I was mixing it with Boeing 747's at San Francisco International and Las Vegas, to clearing dog walkers off the landing strip with low go-arounds at Carmel. Other highlights included flying over the Golden Gate Bridge as the fog moved in, circling the Hoover Dam and exploring the Grand Canyon when it was still legal to fly a single-engine aircraft over the hole in the ground.

By mid-August, I had completed everything I had set out to do. I had the FAA PPL and the grand total of one hundred and thirty hours.

On my return to the UK, I swiftly completed all the necessary written and flying exams until finally, 18 months after I first flew an aircraft, I was the proud owner of an assistant flying instructor rating along with the required Basic Commercial Pilot's License (BCPL). The old adage that you only really start to learn when you're teaching somebody else was to prove 100% correct. Although the pay was this side of pitiful, it was, without doubt, the most rewarding job I ever had within aviation, and to this day, I have kept in contact with a number of my students. I'm pleased to say more than a few have gone on to become airline pilots themselves, many now captains.

It was as an instructor that I first experienced how monetary concerns can quickly get in the way of common sense and the ensuing sometimes tragic consequences. It was a lesson that has still never left me.

I had been appointed to train RAF scholarship students. On the whole, these were pretty switched on individuals and one guy, in particular, had quite obviously had his head in the manuals and was putting his knowledge to good use. I had arrived at the school intending to send him on his first solo. However, the cloud base that afternoon was not great, and in the frequent showers, the visibility dropped considerably. As a result, I informed the cadet that although I had intended to send him solo, the weather was just not good enough so we would practice flap-less 'touch and goes' in the circuit. As we were leaving the office, I bumped into a pleasant middle-aged chap who I had met several times previously. The last time we had chatted, he had told me that he was a private pilot, but he was having difficulty passing his IMC (Instrument) course. As we exchanged pleasantries, he asked if I could help him get his head around a vital part of the IMC course. It concerned tracking to and from beacons. It quickly became apparent that he did not understand even the basic principles and, using his logic, he would fly the aircraft in entirely the opposite direction to that required. I tried my best to explain, but he was having significant difficulty grasping the concepts. The level of his misunderstandings was such that it was not something I could rectify in a two-minute conversation.

He then informed me that the flying school needed a particular aircraft to be flown to Carlisle for an engineering check, and they had offered a 50% discount on the aircraft hire fee if he positioned it over. I suggested he looked out of the window and asked him did he think that the weather was appropriate enough to fly across the country bearing in mind his lack of instrument flying experience and somewhat dubious understanding of the principles behind it. To further hit home the unsuitability of the weather, I explained that I was not going to allow the RAF cadet to fly solo in the circuit. I told him that even with my experience as a pilot and instructor, I would think long and hard about embarking on such a trip in this weather considering the low cloud and possible icing. He explained that he could not afford to turn down flying at a 50% discount and that the school were desperate to have the aircraft in Carlisle by tonight. He then asked me to 'sign him out' for the flight.

During my flying instructor course, I had been given one of the best pieces of advice that have kept my passengers safe over the years. I repeated this advice to the gentleman stood in front of me that day. 'If in doubt, there is no doubt'. I told him I had serious doubt it was safe, and that was the reason I was going to refuse to sign him out.

The IMC course that he had not completed was in effect, an insurance policy. As a private pilot, you must remain clear of cloud at all times unless you have an IMC rating. For any private pilot who unwittingly flies into the cloud without any instrument training, it is very often the last thing they do. That view was confirmed by aviation researchers at the University of Illinois. They took twenty pilots who were only licensed to fly clear of clouds (VMC) and asked them to fly into IMC (cloud) in specially-programmed flight simulators. All of the pilots in the study would have died in a real aircraft. All eventually went into spiral dives that would have ended in uncontrolled flight into terrain or placed the plane in such a state that it would have resulted in structural failure. In repeated tests in the simulator, all of the results were the same. Every pilot lost control of the aircraft. The outcome differed only in the time required before control was lost, which ranged from just 20 seconds to 480 seconds. It appears such a fate befell the helicopter pilot that resulted in the premature deaths of Kobe Bryant and seven other souls in a crash.

After fifteen minutes of flying circuits and touch and goes with the RAF cadet, it was abundantly clear that it would not have been appropriate to send him solo. It was as we were taxiing back to the parking area I noticed that the middle-aged private pilot had not heeded my advice and was taxiing out for take-off. Someone had agreed to sign him out to do the flight. An hour later, he was dead. He had flown lower and lower in an attempt to stay visual with the ground as the cloud base descended. At the same time, the terrain was getting higher. He eventually became disorientated and flew too near a TV mast. A metal support cable sliced through the aircraft's wing giving the poor guy no chance of survival. Of course, I questioned myself. Could I have done more to stop him flying that day? Apart from locking him in the room while I flew the RAF cadet, I don't think I could. He was a private pilot and, ultimately the decision to fly was his, but it was clear that the financial advantage did weigh on his decision-making process of that day and, ultimately, £50 cost him his life.

Within a few weeks of this terrible accident, I was extremely fortunate to be offered a prestigious position as a commercial flying instructor based in Perth, Scotland. It was to be my first salaried flying position as a pilot. Most of the other instructors were either retired airline pilots or ex-senior RAF instructors with a considerable amount of experience, but I was keen, and I was determined. The only fly in this particular ointment was that I needed the full Instrument Rating to take up the post. It was time for another visit to the bank.

I explained to the manager that although I did indeed still have one career development loan outstanding, to stand any chance of paying back the first one, I would need the Instrument Rating. It seemed to do the trick. With credit cards maxed out, career development loan all but spent, it was here that the fifty-pound flight simulator on the Amstrad paid off. I had heard terrifying tales of how expensive the Instrument Rating (IR) course could be. It was, without doubt, the most difficult and most stressful aspect of all the training required to become an airline pilot.

The IR course was to take place in Bournemouth. I had only just arrived at the airport as a student was climbing out of the Aztec twin-engined aircraft.

As he clambered off the wing, in response to me asking 'how's it going', his cheery reply was 'I hope you've brought a razor blade with you for next week.' He was referring to potential hesitation marks on the wrist rather than looking smart for my flight test.

I loved the training and IR flying. Yes, it was, as advertised, very high-pressured and very costly. Still, I had attached the blinkers to these aspects of the course. I worked on the fact that any consideration of the above while flying would only serve to reduce any mental capacity I had left to deal with the task in hand.

It very quickly became apparent that the jerky £50 flight simulator software was probably the best £50 I had ever spent. Within two weeks, I had everything I needed to take up the position of 'Commercial Flying Instructor'.

AST, my new employer, was contracted to train airline pilots for Aer Lingus, Malaysia Airlines, Singapore Airlines and Royal Brunei. The variation in the abilities of the cadets from the various airlines was enormous. The Aer Lingus cadets were there on merit. They were switched on individuals, keen and determined. A significant number of the Malaysian Airline cadets, however, gave the impression that they were doing *me* a big favour by just turning up for lessons. It later transpired that many of these cadets were in fact sons of senior ministers and industrialists in Malaysia. It was no surprise that we were chopping around 50% of the Malaysian cadets before they had even completed the course. AST had dealt with this kind of situation previously. I was reliably informed by one of the senior flight instructors that when a certain middle eastern airline had provided students, the cadets had formed a line in front of the instructors. The senior management of the airline then explicitly pointed out which of the cadets would pass. It just so happened that every one of the golden cadets was related to the royal family. Correctly, the senior trainer at the school politely informed the airline that they would only pass the course if they performed up to the required standard.

Many did not. Many were chopped.

Malaysia Airlines were similarly none too pleased that the school would not rubber-stamp their cadets. At no point was pressure brought to bear by any senior management on any of the instructors to 'overlook' significant failings of these cadets. AST was an organisation with integrity. It also boasted an illustrious list of famous trainee pilots, one of which was the first woman to fly from the UK to Australia, Amy Johnson.

I was more than happy that I was working for such a company.

Unfortunately, six months later, this integrity cost me my dream flying instructor position. Due to the high failure rate, Malaysia Airlines pulled the contract with AST. Within a week of being made redundant, I was contacted by Malaysia Airlines. They enquired would I be interested in training cadets in Malaysia. I asked the obvious question and was reassured that the cadets who had been chopped would not be attending the training in Malaysia. On that basis, I agreed to travel to the other side of the world to take up the position.

I was in a foreign land and back to being paid a pittance to fly. Still, I looked on it as an extended funded holiday in what is now the highly-desirable tropical holiday destination of Langkawi.

As soon as I walked through the doors of the flying school, I recognised several faces. They were the faces of those who had not made the grade in Perth. It appeared that Malaysia Airlines had been somewhat economical with the truth.

Although it was indeed an unforgettable life experience, what I was to witness at the flying school was to open my eyes to *exactly* how corrupt the aviation world can be.

It was my first real encounter with the killer that is the 'Normalisation of Deviance'.

The term was first used by Diane Vaughan in1996 concerning the Challenger shuttle disaster.

*"Social normalisation of deviance means that people within the organisation become so accustomed to a deviation that they don't consider it*

*as deviant, despite the fact that they far exceed their own rules for the elementary safety."*

The question was, when you encounter it, do you condone this behaviour or do you make a stand? My view was then, was the same as it is now. If you don't challenge that behaviour, you are part of the problem. Maybe it came from my punk roots, as Stiff Little Fingers said back in the late '70s. 'Question everything you're told.'

A cadet, who was the son of the Malaysian's finance minister, had already been at the flying school for six months. He was assigned to me as one of my first students when I arrived. The student pilot had already flown solo in the circuit and operated several cross country exercises with an instructor.

After reading his training file, I planned a simple cross country exercise from Langkawi to Alor Star, to Bakersfield, to Penang. There was not a cloud in the sky, and literally, all that was required was to fly east from the island to the coast of peninsular Malaysia, turn right, aim just to the left of the highest mountain around, and then turn right again to Penang. You literally could not miss it.

He did.

The cadet decided the required heading was about 30 degrees to the left of the actual heading needed to fly towards Mount Jerai, the highest mountain in north Malaysia. Despite the fact you could see this mountain from one hundred miles, and despite the hints and all the prompts, he flew towards a ridge of high ground that ran north-south.

This meant having to climb higher and higher as the terrain came up to meet us and the clouds surrounding the peaks got lower. An unplanned flight into IMC is a classic trap that has taken the lives of so many unsuspecting pilots. I have already described one.

More hints, more prompts seemed to harden his resolve to fly in the wrong direction and to press on. I let him get to a state where if I did not take control, it was a certainty that he would have killed both of us.

At this point, I was hoping that the seriousness and severity of his mistake would hit home. If it had, at least he had learned something. Maybe the biggest lesson of his flying life?

I asked him if I had not been there, what did he think would have happened. His response floored me.

'Àllah would take control, inshallah.'

My response. 'Àllah does not have a CPL.'

Another couple of training exercises later confirmed that this was not just a bad day at the office. It resulted in my entering into his training record that this cadet was a severe training risk and his training should be curtailed before he killed himself and before he got the chance to kill others.

Despite this revelation, the chief instructor at the school insisted that I sign the student off to complete his solo cross-country. He explained if I didn't, the school would not get paid by Malaysia Airlines. As I had with the now-deceased aviator from a few months previously, I refused to sign him out. Unlike AST, the integrity of this particular training establishment was highly dubious, and they made it abundantly clear they were not happy about my judgement.

Within a few weeks, we were informed that the Malaysia Airlines training managers were to pay a visit to review how their cadets were progressing. By now, being highly suspicious of the school's view on safety, I was not confident that my review of this particular cadet would remain in the file to be presented to the airline training managers. He was a minister's son after all.

How would I know if the report had been removed and then replaced? The James Bond trick of attaching a hair to the document and the ring binder via sticky tape seemed to be appropriate. This way, I would know if the report had been removed and then replaced. Needless to say, when I next looked at the training file, the hair was snapped.

There followed other numerous examples of downright dangerous practices and attitudes at the school that I was just not prepared to be a party to. The final straw was an incident and subsequent reaction by the head engineer at the flying school that could have easily killed someone.

Part of the checks on most aircraft is to ensure that when commanded, the flaps not only extended to the desired setting but they extend symmetrically. A failure of the flaps to extend is an eventuality that a pilot can be trained to respond to appropriately. Increase the approach and landing speeds and expect a higher nose attitude when landing, and that's it. It is not the end of the world. An asymmetrical extension *would* be the end of the world, at least for the hapless individuals who were in the aircraft at the time. The significant differential in lift produced by the wing with flap compared to the wing with the flap still stowed would immediately result in a violent roll as the plane flipped in the direction of the flapless wing. The pilot would have no time to react. Therefore, asymmetric flap extension, like what to do if a wing falls off, has no place in the training syllabus. If it happens when flying, it's game over.

My student was going through his pre-flight checks, and I was keeping my eye on him, making sure he did not miss anything. As part of the checks, he lowered the flap selector, and the electric motors began to whir. The port flap extended to the commanded state however the starboard flap doggedly remained in the stowed position. I did not intervene. I wanted to see what he would do. He was coming towards the end of his training so very soon he would not have an instructor to hold his hand. The student recycled the lever to try again. This time both extended. What would he do now? Would he move on with the checks and accept the flaps were currently working as they should? He tried them again. They functioned as they should a second time. Would he be happy with that? No, he wasn't. He tried a fourth time. The asymmetry happened again. 'Sir, what do I do?' he asked.

'Assuming I am not here, what *would* you do?' I enquired.

'Get hold of an engineer.'

'Is the right answer' I explained.

I let the student do the talking while I witnessed the interaction between the young cadet pilot in his early twenties and the most senior engineer on the base. The conversation was in Malay. The dismissive way the engineer flicked his hand in front of my student's face and the submissive body language of the young cadet meant I did not need a translator to understand the dynamics of the discussion. The cadet turned towards me. 'He said it would be OK when we get into the air.'

'So what are you going to do about it?'

The poor cadet was in a quandary. He began to stammer. His eyes darted around as he looked for a way out of a predicament not of his making. He had a choice. The first was to accept at face value what the engineer had told him, thereby avoiding a confrontation with an individual that the cadet perceived was more powerful than he was. The downside to avoiding the confrontation was that when he went to extend the flaps for landing, he would have no idea if this would be the last thing on earth he would do.

'Welcome to the world of being a pilot.' I explained.

'So what *do* I do?'

I did not want to tell him what to do, but I wanted to give him the thought process that would save, subsequently, not only his life but the lives of his passengers and crew. I asked a simple question. 'So do you doubt that the flaps will extend correctly?' He nodded.

I recounted the life-saving advice I had been given as a student. 'If in doubt, there is no doubt.' He looked at me blankly. I explained further. 'If you have any doubt about the serviceability of an aircraft, don't accept the aircraft. If you doubt if you are, or you are not fit to fly, don't fly. If you doubt if an aeroplane does, or does not require de-icing, de-ice it. If in doubt, there is no doubt.'

He still did not look happy with the situation. I put it in blunt terms, so he fully understood. 'As a pilot, it is your duty, even as a first officer, to always put safety first. If you are not able to take on board that responsibility, then you don't deserve to be a pilot.'

'So do I tell him I won't accept the aircraft? He will not be pleased.'

'Your job is not to please. Your job is to be safe.'

The engineer was indeed not pleased to be informed he would have to get off his backside and do some work and fix the problem. He began pointing and shouting at the young cadet in an attempt to bully him.

My use of local swear words I had been taught by the cleaner at the school in exchange for teaching Vic Reeves and Bob Mortimer catchphrases seemed to get the message across to the errant engineer. He was in no doubt "I wouldn't let it lie."

The issue is, in middle and far eastern cultures, loss of face is a significant problem with regards to flight safety, as is status when it affects the *command gradient* in the cockpit.

If you ever wondered why the vast proportion of air accidents happen in the middle and far east, now you know.

Within days I resigned and was sitting on a 747 heading back to the UK. I had to wait another six months before that decision was finally to pay dividends.

# CHAPTER FIVE

## *V1 and Bust*

*"'Lessons have been learnt' translates to, 'we got away with it for years until our luck ran out.'"*

Author

In April 1994, now aged thirty-three, I finally managed to get my hands on an airliner. During the six months since my return from Malaysia I had concluded the only way to get a foot in the door, even at a 'third level carrier' was to give prospective employers as few reasons as possible to reject my CV. On that basis, I had taken it upon myself to complete the technical ground school courses and exam for both the Bandeirante and the Shorts 360. I considered this would present one less issue for any potential employer to be concerned about.

The cunning plan worked, and within a couple of weeks, I was sitting in the first officer's seat of a Shorts 360 in Aberdeen. Admittedly it only had thirty-five seats, and it had the looks that only a mother could love. It resembled a flying box with wings. It had no autopilot, it was not pressurised, and it could not fly above ten thousand feet. It was the greatest feeling in the world to operate it.

The company I flew for was called 'Business Air', and it had a regular route for the Shorts 360, which was Aberdeen to Sumburgh. Sumburgh, a place so windy it has no trees and was the only place I have ever flown into that had been closed due to seaweed on the runway.

It was only after gaining around two hundred hours on the Shorts that the next lesson in the realities of working within the aviation industry would be revealed.

It was a winter's morning in 1994 and, as was typical, I was on standby to cover any issues with the airlines main fleet, which were the Saab 340's.

First thing I noticed when I woke up was that overnight it had been snowing, heavily. I was just starting to feel pleased with my lot when the phone rang. It was the airline, and they wanted me in. There had been a last-minute request by an oil company to take workers to Sumburgh, and they needed them there two hours ago.

Upon arriving at the company's office, I walked into the operations room where I would collect and study the paperwork and begin the pre-flight preparations. Even before I received the paperwork, I asked the operations controller 'Has the aircraft been de-iced?' 'Yes, they have done it already,' was his definitive reply.

The problem with ice and aeroplanes is, they don't mix. Even frost on the top of a wing could be enough to prevent the aircraft flying, usually with tragic consequences. There have been numerous crashes due to aeroplanes not being de-iced for various, often ridiculous, reasons. Anything from 'because it will delay a flight' to 'during take-off, the snow will blow off the wings'.

Ten minutes after I arrived, the captain, a young guy from the Welsh valleys, strode in trailing snowy footprints on the carpet. In a thick Welsh accent, the first question he asked as he walked into the crew room was the same as mine to the same person. Only he got an entirely different answer. 'Yes, it is being done now.'

As the captain walked towards me, I told him I had been informed that the aircraft had already been de-iced. Yet now, ten minutes later, the de-icing was 'being done'.

We agreed that we would take great care when we got to the aircraft to make sure that there was no sign of snow or frost on the airframe.

When we did finally board the Shorts, it was completely clear of snow, but not only that, it was also bone dry. If the aircraft had been de-iced, then it should have been dripping in the thick, viscous fluid, sprayed at temperature all over the aircraft. This aircraft had most definitely not seen a de-icing hose in a while. We spoke to the engineer who, within a couple of minutes, had solved this puzzle.

We were informed that the aircraft had landed earlier from a night mail flight and was then towed into the hanger. That would explain why the flight deck had been warmer than expected on this freezing cold morning.

Satisfied with what we could see, or rather not see, and with what the engineer had informed us, we called for the passengers to board the aircraft and began our usual pre-flight checks. I was going to be flying the leg northbound so the captain would be handling the paperwork and the radio.

As soon as we had received push and start clearance from ATC, we started the two Pratt and Whitney engines slung fifteen feet above the tarmac and called for clearance to taxi to the runway.

Although I would be flying the aircraft, the only way to steer the Shorts on the ground was via a steering tiller on the captain's side. As the Celtic captain steered the aircraft along the taxiway, I ran the before take-off checklist, making sure the flaps were at the correct settings, and all the aircraft's somewhat limited avionics tuned to the required frequencies. One last review and we were ready.

The runway in use was RW16, the southerly runway. We would be taking off towards the granite city that sits on a hill.

Having been cleared for take-off, I set the power while steering the aircraft down the centre line of the runway via the rudder pedals. With the power set, I moved my hands away to allow the captain to take control of the levers. Even on the large Airbus and Boeing aircraft, the procedure is precisely the same. Once the take-off power is set, the captain would be the one tasked with aborting a take-off roll if required, therefore until airborne, the thrust or power levers are under his sole and absolute control.

The propellers took large slices out of the dense cold air, as we started to accelerate down the runway. 'Eighty knots'. The call came to make sure both airspeed indicators were working correctly. 'Check' came my response. Almost imperceptibly I was having to keep the aircraft straight as it accelerated down the runway by having to increase the pressure on the right rudder pedal. It can happen if you have a crosswind, especially on such a slab-sided contraption that we were about to go flying in.

When an aircraft takes off, there are a series of V (velocity) speeds that a pilot needs to be aware of. V1 is a speed that, after which, even if an engine fails, you are committed to taking to the air, no matter what. The remaining runway available is insufficient to stop the aircraft safely. Vr or rotate speed is the point where you ease back on the control column to raise the nose of the aircraft to climb into the air. V2 is a speed that if you did have an engine failure would give you the best climb performance. Very suddenly I had to rapidly increase the pressure on the right rudder pedal about ten knots below V1, literally only a couple of seconds before committing to take off. It did not feel right at all. As the pilot flying (PF) it was my job to keep the aircraft straight, steering with my feet, the captain was monitoring the engine instruments. I opened my mouth to call a failure. Before I could get the words out, the Captain called out 'STOP, STOP, ENGINE FAILURE, I HAVE CONTROL'. At the same moment, he slammed the engines into reverse and stamped on the brakes. As per the training, I released the controls over to my Welsh commander and transmitted to Aberdeen tower that we had aborted the take-off. The controller said that the fire crews were already underway as he had seen a large amount of black smoke emanating from the port engine as we neared take off.

Once we had come to a halt, the first thing we needed to assess was, did we have an engine fire? If we did, we would attempt to extinguish it by pulling the relevant fire handle. This would, among other things, cut any remaining fuel to the engine. The rather large hint if we did have an engine fire would be that the fire handle would glow red and the fire alarm would by now be blaring away. Neither was the case.

Having talked it over between ourselves and the fire crew who had arrived in a blaze of sirens and blue flashing lights, we decided to taxi onto the apron with the remaining operational engine so at least we would not block the runway.

We came to a halt, applied the parking brake and shut down the remaining engine on the southern part of Aberdeen Airport's apron.

Within seconds, engineers and firemen descended on the aircraft. One had a step ladder. The height of the engine intake meant this was going to be the only way they were going to get to peer into the offending engine intake.

The first thing that the engineer did was un-lock the cowling that surrounds the engine. Liquid splashed to the floor as the engine cowling was opened. One of the other engineers commented. 'Looks like you had a fuel leak mate.' If the engineer on the ladder had not already got everyone's attention, the next moment, he was guaranteed to be the centre of everyone's focus. 'Bloody hell, look at this!' He reached into the air intake of the engine and scooped out three or four armfuls of partially melted snow that fell to the ground like a deadly slush puppy. The captain and I looked at each other. Expletives abounded.

Already knowing what would be revealed as the cowling was opened on the other engine, we stood there waiting for the inevitable. Five or six armfuls of slush hit the ground.

Thank God the engine failed when it did; seconds later we would have been airborne, taking off towards the granite city. If the donkey had failed after V1, we would have still been committed to take off. With the amount of slush in the other intake, there is little doubt that the other engine would have also failed and require us to turn Union Street in Aberdeen into a runway. In a fight between granite, flesh and aluminium, I would put my money on the granite winning.

I was lucky, very, very lucky. Seconds were the difference between writing this book or my name etched in granite in the granite city. In many ways, those five seconds, or rather the fall out of those five seconds, shaped my entire outlook towards aviation. Simply put, I would not be here now, but for those five seconds.

Immediately we got back to the office we were told to attend a meeting with the director of flight operations (DFO) and chief pilot. It was the moment my *real* aviation experience started.

Very reasonably we were asked to explain what had happened from the moment we had walked into the office that morning. The DFO and chief pilot listened.

'Did you taxi past a snowbank that could have been sucked into the engine intake?'

'No' came my reply. There were no high snow banks at the airport and, unlike most aircraft where the engine intakes are only a few feet off the ground the Shorts engine intakes are much higher.

'Did you check the intakes in your pre-flight checks?' the DFO enquired.

'No' the captain replied.

'Why not?'

Before I realised it, I had answered on the captain's behalf.

'If you don't provide step ladders, how can we check the intakes?'

Here I was, in my naive new boy state, imagining we were going to be congratulated on the way we had handled the emergency. Instead, we were going to be questioned by an interrogator who, it appeared, hadn't even considered the obvious!

The chief pilot and DFO told us the aircraft landed around seven that morning and had been parked outside on the apron for only a short time. They also claimed that immediately after it had been parked, the two engine blanks had been fitted. These blanks are used to plug the air intakes. This is to prevent birds nesting within the intake or to use technical speak 'prevent the ingress of foreign matter,' i.e. birds' nests or snow!

As it had been snowing, the decision had been made to tow the Shorts into the hanger. Allegedly it had stayed there until it was time for us to go flying. According to the chief pilot and DFO, the aircraft had blanks fitted as soon as it had landed. Once the Shorts 360 was towed into the hanger, the snow had then melted off the wings and fuselage.

Given the above information, the DFO and chief pilot's conclusion was that the snow and slush MUST have entered the engine as we taxied out to the runway. Were they were trying to implicate the captain and me?

It was what my father was alluding to all those years ago. 'It's always the pilot's fault.' Who cared if I was the new boy? I was not standing for it! I enquired where the two management pilots believed the engine blanks were stowed when not fitted. The chief pilot informed me that they were kept in the baggage hold on the aircraft. I told them both that in my several months of flying the Shorts I had never once seen the blanks on board, and I most certainly had never seen them fitted. I was already smelling a rat.

The meeting ended pretty soon afterwards. The captain was required to fill in a report, and we completed it together. We highlighted the lack of engine blanks and made sure the CAA knew about it. That was the last we heard about it until a few months later; the investigation report came out with its findings. Despite our claim that there were no blanks on the aircraft, the investigation cited the reason for the ingress of snow was due to the blanks being blown inside the engine intake. This due to strong winds. This then allowed the falling snow to enter the now open intake. So apparently blanks I had never seen were blown into the engine by a strong wind that did not exist. This episode was my introduction to the ways of the airline business. My real *education* was only just beginning.

'Strong winds' were blamed on this incident. Nothing more was done, and there was no further investigation. Could the same thing happen again?

The proof that lessons were not learnt, and our report not taken seriously happened on 27 February 2001. The similarities to my experience were chilling.

A Shorts 360 had landed at Edinburgh Airport just after midnight and was parked on stand in light to moderate snowfall where it remained until later that day. That evening around 17:00 the Shorts had a new crew on board, and they had requested taxi clearance and were instructed to taxi to the north-easterly runway 06.

Unlike our situation, the take-off continued, initially, without incident with climb power selected as the aircraft passed 1200 feet. At 2,200 feet, the aircraft's system that prevents ice build-up was activated. Three seconds later, what had happened to us on the ground happened to this unfortunate crew, however this time, not to one, but both engines. The crew called a MAYDAY and attempted to turn back, eventually ditching in the Firth of Forth. Both pilots died in the crash.

The official report into the tragedy stated that the airline did not have a procedure for flight crews to fit the engine intake blanks in adverse weather conditions. It confirmed that the requirements contained in the aircraft manufacturer's maintenance manual 'Freezing weather-precautions' were not complied with. Furthermore, intake blanks were not provided on the aircraft, nor were any readily available at Edinburgh Airport. The report highlighted that a significant amount of snow had almost certainly entered into the engine air intakes as a result of the aircraft being parked heading directly into strong surface winds during conditions of light to moderate snowfall. Finally, they concluded that the flow characteristics of the engine intake system most probably allowed large volumes of snow, ice or slush to accumulate in areas where it would not have been readily visible to the crew during a routine pre-flight inspection. Does that report ring any bells? Yet again, another fatal accident that could have been avoided because 'lessons had not been learned'. Maybe I had not shouted loudly enough to the CAA to prevent this from happening again, only this time with tragic consequences.

I had experienced the first tragedy that could be attributed to 'tombstone' safety. Tombstone mentality is a pervasive attitude of ignoring design defects or operational deficiencies until people have died because of them.

Within a few months the company I worked for decided that they would retire the Shorts 360 and transfer the crews onto the Saab 340.

Fortunately, at the company I was working for, a new chief pilot named Paul 'Yomper' Yarnold had taken the reins around the same time I was offered my very first command on a commercial aircraft. Every single command decision I ever made at that company was 100% supported by this new chief pilot. I must have made a favourable impact on him and the company. I was soon asked to take on the role of Base Captain at the Glasgow base.

An excellent example of the support I received was the occasion when a cargo door seal had come loose when I landed in Glasgow. As the Saab 340 was a pressurised aircraft, I was concerned that this issue might well lead to a leak when in flight that would require an emergency descent. I rang the operations department to inform them of the situation. I was told by operations that the aircraft was immediately required back at Leeds because it was to be used to fly the chairman of BMI over for a meeting in Europe. Therefore, there was not enough time to have the aircraft fixed in Glasgow. I informed the ops manager firmly, but politely, I would not be leaving Glasgow until it was fixed. The next moment, I could hear the ops manager telling someone of my refusal to fly the aircraft back. The ops manager then went into great detail about the complications and embarrassment this would cause the company concerning the chairman of BMI who, it was reported, was looking to acquire the company. It quickly became apparent the person the ops manager was talking to was Yomper.

Yomper's response was exactly the response any professional airline pilot should expect from his manager.

'Who is it?' Yomper enquired.

'Mike Simkins' came the reply.

'If Mike says he's not flying, then that's good enough for me.'

We subsequently departed about an hour late, and I assume the BMI chairman got to his meeting an hour late. The point is, he got there.

Subsequently, BMI did indeed take over the company I worked for, and it was around this time I was asked if I would be interested in becoming the base captain for the newly opened Leeds base. There was some concern over the number of technical delays that the Leeds base was experiencing. As a result, I was also invited to become the pilot representative on an engineering task force to look into the reasons.

It was crystal clear what was behind these issues, and there was a straightforward fix.

It transpired, defects were not being entered in the tech log, but passed by word-of-mouth to the oncoming crew. This meant no one had any idea when the defect first occurred, and, as is the way with defects, they only get worse. At some point, the component suffering the defect would finally, totally fail. Sod's law would dictate this would happen at the furthest flung airfields we flew to.

All it took was a week or so for the word to get around that it was not a good idea to greet me with 'we had a problem but, so you know I've not put it in the tech log'. My response would be that I would not accept the aircraft until they did, spiced with a few added expletives.

This did not make me many friends within the engineering fraternity nor, initially, with the other pilots at the base, but within only a few weeks, this change in culture began to bear fruit. More to the point, it highlighted an issue that could have very well become a killer.

There had been a recurring problem with one of the Leeds based aircraft. The Saab's rudder would jam. It was continually put down to an electrical fault.

One day I was due to take this particular aircraft from Leeds to Paris. As part of my pre-flight duties, I had a look at the tech log and found that in the previous week there had been two issues when landing where the captain had noticed the rudder was severely limited in movement. On both occasions, this was put down to the failure of the rudder limiter. Again, both times, the aircraft was allowed to fly back by isolating the rudder limiter.

As we landed back into Leeds, it became apparent that once again, the rudder would not move. Once the passengers were off the aircraft, the engineer came on board to enquire whether the aircraft was serviceable. I told him about the restricted rudder movement 'oh yes we know about that I'll just reset the system' was his reply.

I informed the engineer neither the aircraft nor I were going anywhere until the reason for this recurring issue was finally isolated.

Once the control cable runs that convey the movement of the rudder pedals to the rudder were checked, it became abundantly clear where the issue lay. There had been a directive from the manufacturer Saab several years earlier of occurrences of ice build-up on those control runs. To prevent this from happening, warm air from the engine was to be diverted into the tail-cone. The two clumps of ice on the control cables bore witness that this directive had not been complied with. Fortunately, on that flight, I had not experienced a crosswind landing at Leeds with the limited rudder movement. Not having to contend with a strong crosswind at Leeds in itself was something of a rarity. The consequences, however of an engine failure with the same issue did not bear thinking about.

A strong crosswind invited the next insight into the workings of airline management thinking.

It was only days after I had overseen the engineer's discovery of the two clumps of ice on the rudder cables that I walked into the crew room to be greeted by two white-faced pilots.

These particular pilots were supposed to be, at this point, on the return leg from Glasgow.

'What are you guys still doing here?' I asked.

The captain replied, 'I take it you've not looked outside yet?'

It transpired that the aircraft had never left Leeds. During the take-off run, the plane was blown off the runway. It was now facing the opposite direction, splattered with mud with its wheels stuck into the ground. It was the same aircraft upon which I had highlighted the rudder issue.

Within minutes a non-flying, ex cabin crew, base manager for BMI wanted to have a word with me. She informed me that, it had come from on high that if I was approached by any members of the press, as the base captain, I was to deny it was a BMI flight. I was to inform them that it was operated by Business Air. This was a company that had ceased to exist around six months previously. I pointed out to her the fact that the aircraft, turned tractor, had the BMI colour scheme emblazoned on the fuselage, the passenger tickets said it was a BMI flight, with a BMI flight number. It had BMI cabin crew, and we were employed by BMI. This seemed to make little difference to her argument. I then highlighted the apparent fact that the company she wanted to blame no longer existed might make it quite easy for any journalist who did not write with a crayon to see through the crass PR attempt at subterfuge. She was, however, insistent. I was equally emphatic that if BMI wanted that outright dishonest statement to perpetuate, then they better hope that I was not interviewed.

# CHAPTER SIX

## *Common Aeronautical Language C'est La Vie.*

*"Common Aeronautical Language c'est la vie."*
CHIRPS Publication (Confidential reporting system For Aviation)

An incident I experienced first-hand during my last few months as base captain at Leeds highlighted that it was not just commercial implications that trumped flight safety. Another tragic accident exposed that misplaced national pride was just as lethal a cocktail to passengers and crew.

The tragedy that follows, when considered alongside the fatal crash of the Shorts 360 at Edinburgh, are ultimately the reasons that reinforced my uncompromising stand this book later describes.

This would become the second time that the authorities ignored my precise and direct highlighting of an obvious risk to flight safety. It was the second time that a fatal air crash occurred as a result.

In this sorry tale, it was not just the UK regulator, the CAA, who did not take the appropriate action. It was also the French regulator, the Direction générale de l'aviation civile (DGAC) and European court and unions.

In this particular episode, it appeared the nationalistic tendencies of the French were to have tragic consequences.

With the Charles De Gaulle (CDG) route being a regular destination from Leeds, I started to pick up a pattern in the order that aircraft were being cleared to take off at France's busiest airport. What is supposed to happen, subject to any air traffic slots, is that aircraft are cleared to take off in the order that they arrive at the holding points for the departure runway.

It became apparent, however, that the "first come first served approach" was not the case at CDG. National favouritism was in full effect with not even a hint of concealment. It was so blatant that I found myself admiring them.

The process would go as follows. A French aircraft, be it a short haul or long haul aircraft would taxi all the way to the holding point at the runway end. Every other, non-French aircraft, would be cleared to taxi for an *intersection* departure. The tragedy that I describe later was just but one example of this practice at CDG.

The CDG plan then became clear. There would be, in effect, two queues of aircraft split along nationalistic lines; French, and then the rest of the world. Irrespective of the order that the aircraft arrived at the holding point, the take-off order would invariably be an Air France aircraft at the runway end, then a non-French aircraft from the intersection, followed by another Air France aircraft from runway end, and so on.

It was only after the numerous times that I witnessed an Air France aircraft cleared to take off before my aircraft, even though I had been waiting longer at the holding point that I decided to do something about it. I explained to my first officer my plan of action that would save my company thousands of pounds over the year.

'Midland 498 request taxi', the CDG ground controller cleared us to taxi to the usual intersection departure. 'Negative, we require the full length for departure. Midland 498.' The controller, as I had fully expected, questioned, why we needed all the runway length. I informed him that as the captain of the aircraft, I could have any length of runway that I deemed necessary, and I wanted the full length.

The fact that there was now a BMI aircraft in the queue at the holding point that was reserved exclusively for Air France had, to put it mildly, put a rather large Saab 340 shaped fly in the ATC ointment.

The cunning plan worked, and we were cleared to take off a good ten to fifteen minutes ahead of the usual intersection departure. The money BMI saved in fuel burnt sitting at holding points and the fact that the passengers would arrive back at Leeds on time made the effort worthwhile.

We had put French air traffic in a bind. They had to get us out of the way to get the Air France flow going again. It became the standard practice after that to be sitting at the holding point as the filling in the sandwich between, for example, an Air France Concorde and an Air France 747. On one of these occasions an Air France Concorde captain who knew exactly what we were up to transmitted 'I like your style Midland', as we taxied onto the runway ahead of the magnificent symbol of what we used to be able to design.

A year or so later saw the demise of this beautiful aircraft. According to the authority's version of events, the crash was caused by debris on the runway. This was despite two firemen who were eyewitness to the crash, stating on oath, that they saw Concorde on fire before the alleged position of the debris on the runway. The official report pointed to an alloy strip that had fallen off a Continental Airlines DC10 departing for Newark as causing the crash. I would submit the 'official' reason for the crash is substantially at odds with the facts. A spacer had been omitted from being fitted to the undercarriage. Many believe, including Concord pilots, that was the reason the tyres exploded.

What I can confirm is, this was not the first time that debris had been on the runway with an aircraft cleared to take off at CDG. Two years prior to the crash, I witnessed an Alitalia aircraft at the holding point at CDG waiting for line up clearance. There was an aircraft about to land, however on his roll-out, all was not well. The pilot reported that he had blown a tire on landing and suspected it had resulted in debris on the runway. The aircraft was not French; he gave the information in English and in English came the reply.

With that, the female controller acknowledged the transmission and then immediately gave clearance for the Alitalia aircraft to line up and take off. This was met with a somewhat disgruntled reply from the Alitalia pilot. 'We will hold the position; didn't the previous landing aircraft just report debris on the runway?'

ATC replied, sounding somewhat surprised 'confirm you wish to hold?'

Imagine for a moment if the landing aircraft had been French, and ATC and pilot had talked about the debris in French. The Alitalia may have accepted the clearance with possibly disastrous consequences. But of course, all ATC transmissions are in English for that very reason – air safety.

Or are they?

It had become apparent on my flights into CDG that when entering French airspace, any French registered aircraft would receive and reply to any air traffic instructions in French. I had always been under the impression that air traffic control (ATC), no matter where in the world, would still give ATC instructions in English. The reason is simple. It enhances flight safety. As a pilot, it allows you to pick up subtle clues about what other aircraft are doing by listening to all the communication on any particular frequency. It helps aircrew to build a mental picture of the environment and traffic direction in your specific airspace. There is a phrase for it. "Spatial awareness."

I had flown into and out of CDG several times from the Leeds base, and I had formed the opinion that the standard of air traffic control in the airspace around the eighth busiest airport in the world, left a lot to be desired. For example, we would receive lots of last-minute changes to arrival procedures that would significantly increase the workload. On a couple of occasions when thunderstorms were in the area, the controller would become so quickly overloaded they would simply not respond to requests to change headings to avoid the cells. At that point, you chose to fly through a thunderstorm and risk severe damage to the aircraft and probably passengers, or, you took it upon yourself to change heading without a clearance. I'm still here, so I leave it to you to consider what I did.

All these issues were further compounded by the fact that, as a large number of aircraft communicated in French, there was no way that it was possible to build up any relevant internal spatial awareness.

One particular evening I was rostered to fly into CDG, and all was going to plan until the late stages of the approach. The number of ATC transmissions was high, as it was the peak time. Aircraft from different departure points throughout the world were converging on the French capital's main airport. As usual, there was a high amount of ATC transmissions being conducted in French. We had been given a heading by the CDG radar controller to intercept the final approach to the easterly runway. We began to slow down to configure the aircraft to make the landing. Once established on the inbound path to the runway, we were instructed to contact CDG tower for landing clearance. Again the amount of chatter on the frequency was so high that we had difficulty in actually informing the Tower that we were on the final approach. Finally, there was silence, and we were able to transmit,

'British Midland 497 localiser established.'

The Tower replied 'Midland 497 cleared to land' along with, as is standard, the wind speed at that moment. However, there was one small problem, well rather a significant, big, metal problem. Still on the runway was a Boeing 747. Fortunately, the visibility was excellent, and we could see the aircraft clearly, and it's very slow progress to the runway turnoff. We continued down the ILS in the hope that the other aircraft would clear the runway in time. The ILS consists of what in simple terms are two radio beams that receivers pick up in the aeroplane that will give both glide slope and lateral displacement information to the flight crew. That is how a modern airliner can land even in low cloud or low visibility.

It became evident that the preceding aircraft would not clear the runway in time, and I decided to commence a missed approach as this would be the safest option. This manoeuvre happens reasonably regularly and can be caused by bad weather, maybe strong winds, turbulence at a low altitude, more substantial than forecast tailwinds or, as on this occasion, where the flight crew decided it was the safest course of action. It's a flight manoeuvre that is practised repeatedly in the simulator. Before every approach commences, the pilots brief on what to do in the event of a go-around. Despite the sensational headlines that appear in newspapers, a go-around is a non-event. However, the reason for the go-around might be either mundane, as above, or something far more serious.

On this particular occasion, we flew the missed approach procedure, with RT (radio transmission) to the tower mixed with expletives in the cockpit as to why the controller had cleared us to land when the runway was not clear.

We flew the proscribed headings and altitudes as per the published procedure and then, again, received headings and altitudes to guide us back on to the final approach and ILS.

The constant RT chatter, which was again liberally sprinkled with French, made it impossible to inform CDG tower that once again, we were at a late stage of the approach. Finally, we managed to get a word in edgeways, and we were given the clearance to land. This time it was without several tonnes of metal still occupying the runway. The rest of the flight and the flight back to Leeds continued without further incident.

There, however, *was* a problem, and it was a problem that needed to be addressed. Anybody with any sense could see it; even people with no flying experience could see it.

As a pilot, you must highlight any weakness in procedure or equipment, or indeed yourself, if for no other reason than self-preservation. Having discussed what had happened with the first officer, we decided to complete the form designed to highlight any such weakness. The form is initially submitted to the safety office within an airline. If the safety office determines that the problem highlighted is a recurring problem, or is of such seriousness that it requires immediate action be taken, then the form is sent directly to the controlling authority which, in the UK is the CAA.

I completed the required form and indicated that, in my assessment, there were two reasons for the go-around.

Firstly, the indisputable fact was that there was another aircraft on the runway when we were cleared to land. In my opinion, this would have made any attempt to land unsafe. Secondly, the number of communications in French at CDG, mixed with a large amount of RT (Radio Transmission), chatter made it virtually impossible to build sufficiently robust situational awareness at an airfield that has been established as being one of the busiest in the world.

We placed the completed form in the post-flight envelope and that, I thought, was that. What is important to realise is that the details contained in this form are not designed for public consumption; it is an internal means of dialogue between the aircrew, their respective company and, in this case, the CAA.

The Leeds base was extremely sociable, and there would be regular evenings out with the pilots, cabin crew and wives/husbands etc. It was at one particular base curry night that a colleague asked me, "Hey did you see this month's PILOT magazine? You're famous!".

The very magazine I flicked through years previously now described what had happened at CDG. I pointed out to the bearer of the news that I had never actually sent any articles to the publication. My colleague continued to inform me that the story related in detail, not only to the go-around itself, but also the reasons behind the go-around. I thought this was a bizarre occurrence. The only information of the incident was contained in the confidential form I had completed. So, how exactly did PILOT magazine get hold of the story?

The following morning I visited the local newsagents and seeing there was no-one around, I reached for the PILOT magazine on the shelf. Eventually, I found the page. It was virtually, word for word what I had written in the confidential correspondence to my company's safety department. The details had, however, been de-identified. But there, in black and white, were the exact details of what had happened. Someone somewhere had leaked the information.

Within a few short weeks, the article appeared in CHIRPS. The Confidential Human Factors Incident Reporting Programme is a publication that provides an independent and confidential reporting system for all individuals employed within the aviation industry.

Again, this publication is reliant upon individuals supplying reports directly to them. The information is de-identified and published if the board consider that the incident needs highlighting to the broader aviation community. You can, therefore, imagine my surprise when through my letterbox came the latest edition of CHIRP with the headline "Common Aeronautical Language C'est la vie!" Containing my report in full. As with PILOT magazine, I hadn't sent CHIRPS the report.

From Issue 48 From October 1998, here is exactly what appears.

"The following text is taken from the CAA (SRG) Occurrence List (ATC) 1st Jun 98 – 30th June 98 and merits particular attention".

OCC NUM 980151QJ

"Pilot states that during approach, non–standard R/T (French) was used and that his aircraft was on very short finals without having a clearance to land. A previous a/c had landed on the r/w and called clear, but was not. Go around flown. On second approach, the crew again found it difficult to call "LOC established", due to continuous radio transmissions in French.

CAA CLOSURE: The French Authority's report indicates that the reporters a/c failed to follow a reduce speed instruction and this, combined with the previous a/c's slow clearance from the r/w led to ATC instructing a missed approach."

The final paragraph is the *killer.*

Concerning the use of French, the report stated, "English and French are used jointly at Paris Charles de Gaulle in conformity to the

MINISTERIAL ORDER of 7th SEPTEMBER 1984 relating to radiotelephony procedures for the use of general air traffic. It states in particular (in paragraph 2.4) that French must, except in special case, be used between French flying personnel and French ground stations."

There were two problems with that French reply. The first was, having discussed the matter with the first officer, neither of us remembers being given an instruction to slow down and secondly, *we* had decided to go around, not French ATC. More tellingly and maybe an insight into the shenanigans within the industry, the details of the flight such as flight number, date, and airline information were not passed to the French authorities. It is a confidential process with the information being "de-identified". Therefore, how would they know if a speed reduction request was or was not given if they have no idea when this situation occurred, our call-sign, or the time or even which airport it happened? It wouldn't be the French authorities trying to cover up, would it?

Well, let's see.

A few days prior to the CHIRP publication, I had been involved in another situation at CDG that once again was directly attributed to ATC using French at this major airport. This time the incident could have been disastrous.

On the day in question, the visibility was poor, at only 2000 meters, and as usual, there was a mix of French and English being spoken. These conditions are exactly why one language should be used to help pilots maintain their spatial awareness.

A Boeing 757 was lined up on the runway and cleared to take off. We were 'cleared to line up behind the departing 757' who was only now spooling up his engines. Again there were several transmissions in French. Immediately the wheels of the 757 left the ground we were cleared for 'immediate take-off'.

Our reply was clear and concise, 'negative we require three minutes for wake separation.' This was non-negotiable. It's internationally mandated, and the French ATC knew this. Once more, there were ATC transmissions made in French, and even though we could not understand what was said, the stress in the controller's voice was tangible.

'I say again you are cleared IMMEDIATE take off' barked ATC.

I again refused the clearance.

'You must take off IMMEDIATELY'.

'NEGATIVE' my first officer replied.

If a large aircraft takes off in front of a smaller plane, the ATC unit must give at least two minutes separation between the larger aircraft and the smaller one. The period increases to three minutes if the smaller aircraft is departing from a different position than the larger aircraft. Wake vortex turbulence can best be described as the turbulent airflow that forms behind an aeroplane as it passes through the air. It begins from the moment the proceeding aircraft rotates skywards.

The reason the tower controller wanted us to risk taking off and ignore the years of study into aircraft accidents caused by wake turbulence was this. The transmissions on the ATC frequency were actually from an Air France aircraft that was on final approach to the runway we were refusing to take off from. The controller was caught out. He should have waited for the Air France aircraft to land first, before lining us up on the runway behind the 757. He then compounded his mistake by asking us to risk a take-off into a known danger. What he *should* have done, once he realised his mistake, as we were sitting on the runway, was to instruct the Air France aircraft to go around. Sensing what was developing, I had started to move my aircraft forward.

'You vacate the runway immediately.'

This instruction I could follow, and we taxied off the runway. Moments later, the Air France aircraft landed. Once I was sure that he had landed and before anyone else had been cleared to take off, I informed the controller, over the air, that due to his incompetence, he had caused a potentially dangerous situation to which I would be reporting fully to the relevant authorities.

I filed a report on the occurrence. However, when I read the CHIRP publication and how the French authorities responded to the first incident, I decided to also write directly to CHIRP. Not only did I answer the reply given by the French authorities to the initial story, I also informed them about the latest incident.

I also made a chilling prophecy.

Again this is an exact copy of what appeared in CHIRP. Note the date:

A Common Aeronautical Language (FB48) Jan 1999.

"I was the captain in this incident. The French ATC would appear, according to the French authorities, to be blameless in this incident at Charles De Gaulle ... Well, let's see: We DID NOT fail to follow a reduced speed instruction; at least one given in English. Maybe one was given in French? It has happened on more than one occasion when things get fraught at CDG – the local controllers revert to native language. A further incident may illustrate the problem and, I am sad to say, maybe show how a major incident is waiting to happen if the French ATC continue to be constrained by Ministerial Order No 7.

The visibility at the time was 2,000m and low cloud base. As usual at CDG, chatter was heard in French and English. A B757 (UK airline) was cleared for take-off ahead of us, we were immediately given clearance to line-up, followed by immediate take-off clearance. The first officer informed Tower that we needed the standard separation for wake vortex. Again more chatter was heard in French. We were then requested to vacate the runway as an Air France B737 was approaching on very short finals. This raises the point that had the transmission been made in English, as a crew, we would have been able to build up the 'mental picture' of the situation. In this instance, knowing we would have to wait on the runway, we would have declined the clearance to line-up knowing the B737 aircraft was on short finals. As it is, the simple fact that French is used in RTF transmissions has cost our company fuel for go-round in the first incident and more taxi fuel and an extra 20-minute delay in the second. More importantly, the use of two languages has caused flight safety to be eroded. Is this a case of nationalism over safety?"

CHIRPS Reply

"We have received other similar comments on this matter and will seek the support of CAA to once again represent these concerns in the appropriate forum. It is important to report formally every occurrence of this type that prejudices safety."

There was obviously a problem at CDG, and I decided to do whatever I could in an attempt to get it resolved.

As the base captain at Leeds, I had asked all my crew flying into CDG that every time French was spoken on the RT to put in a report as it was a flight safety issue. We had two flights a day, so within a couple of weeks, there were over twenty flight reports all sent to head office, all saying pretty much the same thing. A few days later came the call. It was my fleet manager enquiring why he had received such a large number of reports from my base. I told him about the two incidents I had been involved in and that it was only a matter of time before there was a disaster. I also told him that when there was an accident, if asked about it, I would have to say that my base had informed the company over twenty times that this was an accident waiting to happen, and it was one that could be easily rectified. The fleet manager then told me he would forward all the details to the CAA.

A few weeks later, I received a phone call from a senior CAA official. Again, like my fleet manager, he was interested why I had asked my colleagues to go to the trouble of filling in a report every time French was used at CDG. I gave him the exact same reason I had given my fleet manager.

It took less than a year for this prediction to manifest itself with tragic consequences.

You may have never heard about the accident. It is unsurprising that the authorities, naturally, wished to keep the lid on this news story due to the embarrassment to the CAA, the French authorities, and not least to French ATC and its government.

The report was published in Flight International, a weekly publication for professional pilots. The headline was "Language Confusion at CDG Caused Collision Claims BEA."

The magazine ran a report on the accident by the Bureau Enquêtes-Accidents (BEA). BEA is the French authority responsible for safety investigations into accidents or incidents in civil aviation. The weakness of the BEA, is, as it states on its website homepage "The identification of causes does not imply the assignment of fault or the determination of administrative, civil or criminal liability." Some might, therefore, conclude it is, in fact, a toothless organisation.

It was dark when an Air Liberté MD 83, began to taxi from terminal 1 to the holding point of runway 27. The aircraft stopped on the way to the runway to deal with a technical problem. A British Shorts 330, was cleared to taxi from the cargo ramp to runway 27. At this point, both aircraft were under the guidance of a ground controller.

The British aircraft was cleared to an intermediate take-off point on runway 27. Shortly afterwards, the Air Liberté aircraft resumed taxiing for a full length departure.

The Air Liberté was then told to line-up and wait on the runway after a landing B737, but he was told in French.

As the B737 vacated the runway, having passed in front of the Shorts 330, the Air Liberté was then cleared for take-off. Again in French.

Five seconds later, at the intersection, the British aircraft was 'cleared to line up on runway 27 and wait' but in English. The British aircraft taxied forward and entered the runway. Shortly before impact, the Shorts 330 captain noticed the Air Liberté beacon lights and braked. About the same time, the Air Liberté crew noticed the Shorts 330 on the edge of the runway. The aircraft had by then passed V1, a speed past which the plane is committed to getting airborne.

The left wing of the Air Liberté collided with the right propeller and cut through the Shorts 330 cockpit, decapitating the first officer of the British aircraft.

Due to the French denial that speaking two languages on ATC is inherently dangerous, there was now a man dead. The only comfort about that evening was that only one person died, not that this will be of any comfort to the unfortunate pilot's family. It could very easily have been six or seven hundred deaths if it had been two wide-bodied jets colliding instead of a small mail aircraft.

The whole situation is a little like knowing that there is a bomb that has been placed at Charles De Gaul Airport. Not only do we know that there is a bomb, but we know exactly where the bomb has been placed. The only things we don't know is, how big the bomb is, and when it will go off. That, however, is immaterial because if you move or defuse the bomb before it explodes, it will not kill anyone. The French and the European air traffic agencies don't want to move or defuse this particular bomb. The bomb is still ticking at CDG airport. As in any crash, there inevitably will be more than one cause, as was the case in this one. However, the findings, by the BEA, the French authority responsible for safety investigations into accidents or incidents make for fascinating reading.

The agency responsible for all air traffic movements throughout Europe is Euro Control.

Euro Control has a publication, dedicated to the role of ATC in air traffic safety; its name is "HindSight" with the subheading "Putting Safety First In Air Traffic Control."

It was in HindSight that the BEA investigation was reported. They gave a significant part of their January 2005 edition over to the Shorts 330 crash.

So what did "HindSight" have to say about the crash? In its "analysis section" the first thing it states is "All the exchanges between ATC and the French aircraft were conducted in French while all communication with the British aircraft were in English." However, when it came to the "probable cause" of the accident, it was nowhere to be seen. No link made. Bear in mind this information was taken from the official French BEA findings.

The "probable causes" given were;

a) The Tower controller's erroneous perception of the position of the aircraft.

b) The inadequacy of the systematic verification procedures at CDG, which made it impossible for the errors to be corrected.

c) The Shorts crew not dispelling any doubts they had as to the position of the other aircraft before entering the runway.

If English had been used, it is unlikely the last two errors would have occurred. It was only towards the end of the article, within the contributory factors, was it mentioned that the use of two languages meant the Shorts crew were not conscious that the MD83 was going to take off.

In the "recommendations" section it suggested the French authorities study the suitability and methods of implementation for the use of the English language for air traffic control at Paris Charles de Gaulle aerodrome. They also considered the extension of this measure to other aerodromes with significant international traffic. In making this recommendation, it was stressed that the investigation did not aim to evaluate the advantages and disadvantages of the systematic use of a single language.

The French said 'Non' to that recommendation.

French is still used at CDG. The bomb is still sitting there ticking, tick, tock.

# CHAPTER SEVEN

## *Living the Dream*

*"Don't discuss anything that can be explained on the ground or is not pertinent to the operation of the aircraft at that time."*

Airtours Command Course Examiner

Do I stick or do I twist?

Working as a base captain for an airline at Leeds Airport far exceeded any position I could have dreamt of attaining only a few short years previously. Despite the late start to my career, I had somehow found myself overseeing twenty or so professional pilots flying an advanced passenger aircraft.

It wasn't a jet.

One of the main restrictions on pilots moving from one airline to another is that of 'seniority'. If I did apply to the jet operators, I would start right at the bottom of the seniority ladder with no job security and, in effect, the years spent as a turboprop captain wasted.

Big decision.

I decided to throw my hat into the ring and apply to one of these operators. I had heard excellent reports from pilots about a company called Airtours. The time to command was around five years.

Within a few short weeks, I attended an interview at the headquarters of the jet operator. The meeting was conducted by the head of training and the director of flight operations. It was apparent that these individuals were not so much interested in my flying or technical ability but were concentrating on the more fundamental aspect of 'could I spend more than a couple of hours with this individual in a cockpit without wanting to throttle him?'

The interview had gone very well, and, by the end of October 1998, I had begun my training to fly the Airbus 320. I had finally reached my goal of operating a large passenger jet aircraft.

In the intervening period, my daughter Katie was born. I was now more concerned about mastering the technique behind changing a baby's nappy than concerns over how a fly by wire aircraft was controlled. I was not even sure which one was the more difficult task to master.

The time between handing in my notice at BMI and beginning the ground school studying the aircraft systems was not wasted. Not only did I become a dab hand at changing baby Katie, but I had also got my hands on as much pre-course learning material as I could. The result was, by the time I was reporting for ground school, I already knew the systems.

After nine months at the company, I was sent to Toulouse to complete the training for the magnificent Airbus 330. With the Airbus 330 rating, I began to fly long haul routes as a first officer.

I was finally living the dream at thirty-eight years of age, ten years after I first began my flight training!

My very first long haul training sector was with the Airbus fleet manager. It was a flight to Jamaica. That evening, over a beer, he suggested that I should keep my head in the books and remarked that the chief pilot had given him very positive feedback on a previous trip that we had flown together.

'Who knows Mike, this time next year you might be in the left-hand seat.' I attempted to downplay this expectation by reminding him I was one hundred and twenty-six on the first officer's seniority list. After a couple of sips of beer, he smiled and said seniority was just one of the criteria that they assessed for commands.

Eighteen months after joining Airtours, a letter arrived informing me I was to be offered, subject to passing the training, a position as a captain.

I had heard several rumours that Airtours had done backroom deals with turboprop operators like BMI Commuter, in that they would not take their more senior captains until the smaller outfit could handle the turnover of aircrew. It was to ensure there would always be a certain level of experience at the third level carriers to ensure passenger safety. The fact was, I was now being invited to commence a command training course, even though I was one hundred and twenty-six on the first officer seniority list. Although I had no proof, this suggested to me that my rapid promotion was all part of a backroom deal. I just wished

I had known about this in advance as it would have made life a lot easier for the two years I had been kept waiting for a commencement date with Airtours after my interview.

The command course went like a dream. It actually could not have gone any better.

I was paired off with another would-be captain by the name Tom Mackle. Tom has since done very well for himself becoming the head of DHLs' aviation division.

We had both prepared for this moment, knowing that this was an opportunity of a lifetime. Manuals devoured, checklists memorised, and strategies for virtually every eventuality designed. For weeks I had pestered every captain I had flown with since being informed of the course. They must have got sick and tired of my 'what would you do if' questions.

After the first couple of days, Tom and I had convinced the training captain we were competent and confident handling the airbus. It was at this point he started to probe to see if *the decision-making process was what would be expected from a commander.*

Two examples of this were during the final test. The airbus is fitted with a device known as TCAS. This stands for Traffic Collision Avoidance System. In the event of another aircraft becoming a threat, initially, the system will show the other aircraft as a hollow diamond on both pilots' navigation display. It changes to a white diamond when the other aircraft approaches within one thousand feet. If the target gets even closer, this changes to an amber circle with the system now announcing 'TRAFFIC'. If it really is not your day and it becomes a collision risk, the target becomes a red square. The vertical speed indicator (VSI) turns red and green. An automated voice states, depending on various factors 'TRAFFIC - CLIMB' or 'TRAFFIC - DESCEND'. Keep the climb, or decent in the green band on the VSI and you will miss the other aircraft. Simple?

Here is a test. You have a TCAS blaring away 'TRAFFIC - DESCEND'. At the very same time, air traffic control is screaming at you to climb due to an aircraft on a collision course. What do you do? Come on! You have literally a few seconds to make this life or death call. Heads or tails? Climb or decent? Live or Die?

Have you chosen to obey the human ATC direction?

So did a Russian aircraft. It resulted in a mid-air collision with a DHL cargo aircraft 36,000ft above Uberlingen in 2002.

You must always obey the TCAS.

Now imagine dealing with that situation while also being given a warning that your generator is overheating. What do you do first? Fly the aircraft or deal with something that potentially can cause a fire?

Welcome to the last day of a command assessment.

During the assessment, and to add to the pressure, a further scenario was given. I was told this situation might occur by various captains.

We were given an issue that required us to divert. The first officer became involved in a conversation about the why's and wherefores of my decision to divert with the examiner who was by now playing the role of the company's operations department.

After several attempts to get the first officer off the ops radio frequency, I eventually had to up the heat. 'I have made my decision. Tell them we will speak to them on the ground in Gatwick and get off the radio!'

After the exam, during the debrief, the examiner stated that was exactly the right thing to do in that situation. 'Don't discuss anything that can be explained on the ground or is not pertinent to the operation of the aircraft at that time.' The management pilot who was witnessing the assessment agreed that the flight, and scenarios, were handled correctly.

I passed the test with an excellent write-up. Two hugely-experienced training captains, one who was the fleet manager for the Airbus, had both concluded that *'my decision-making process was what would be expected from a commander'* .

Remember that phrase.

The next stage was line training. This is a series of flights with fare-paying passengers, mentored by an experienced training captain.

The morning of my first line training sector broke. I was all set, bright-eyed and bushy-tailed. Unfortunately, the company had double-booked the training captain. The same thing happened the next day. The third day the telephone rang. The conversation that followed convinced me I had made the right move by joining Airtours.

It was the chief pilot Jack Brown. Jack would call a JCB a JCB. Everyone knew where they stood with Jack, and he swore like a trooper. He was a pilot's chief pilot, who had previously flown the BAC Lightning in the RAF, all when I was still in short trousers.

'Mike! Hi, it's Jack, sorry about this, but we have fucked up.' Jack went on to explain that the company had yet again mixed up the training slot and because of the delay and that the next month's training schedule had already been assigned, it would be a full three months before I could expect to begin my line training.

'I bet you're a bit pissed off with us, aren't you?'

'Well if you think you are getting a Christmas card this year, you better think again,' I informed him.

'So you won't want to hear that I am going to backdate your salary increase to the date you would have passed when you do finally complete your line sectors?'

'I shall put your card in the post right now, so you get it well before the Christmas rush.'

Yes, it was the company's error, but here was the chief pilot apologising, even though it was not his mistake. Not only that, he was sufficiently generous to ensure that I did not lose out financially due to someone else's repeated failure. This was precisely the type of company with whom I wanted to be associated. I was proud to fly for this airline!

The senior pilot management within the airline were straight-talking, honest individuals with integrity. What more could any pilot want?

From the top-down, from DFO, chief pilot, chief training captain, to the base pilot managers, the airline had integrity, experience, common sense, and honesty. It looked after its crew. Of course, it was not perfect; no company ever can be, but if you had an issue, there was direct access to the senior managers. You might not even agree on a particular problem, but you were listened to, and at no point during those years did I feel any urge to leave and join the 'big two,' - British Airways or Virgin. I had found 'my airline'.

I was now flying all over the world as commander on the Airbus 330. I was also flying short-haul on the Airbus 320 and 321. To top everything off, I also became a father again, this time to a small bundle of joy called Jake.

It was during this period that a first officer asked me a question to which I gave a prophetic answer. British Airways were heavily recruiting at the time. He asked if I was going to apply. My response? Why would I want to join a company that had a rigorous hierarchy and, as a first officer, no real possibility of ever getting a command before retirement? I was flying a modern aircraft all over the globe, I was listened to by the management, and more importantly, I trusted them, and they trusted me. Although they might question a decision I made, they would always support it after I explained my rationale.

I added almost as a throwaway comment that. 'I shall be here until I retire unless I fall out with someone.'

# CHAPTER EIGHT

## *A Bucket and Spade Messiah*

*"Don't make a decision until you have given them options."*

Capt. John Smith, Business Air

One constant in the airline business is, nothing is ever constant. Personnel change, senior managers retire, and of course, the industry is always in a state of flux with airlines coming and going. Companies merge to form more prominent companies and, sometimes airlines fail. As a consequence of any of the former, the culture within that airline may change and not always for the better.

For at least ten years there had been give and take on both sides. Of course, issues arose from time to time but at no point did I ever personally feel my career prospects would ever be damaged by raising any safety or legality concerns. In the days of Jack Brown, the very few times I had, these were dealt with promptly and on a few occasions even resulted in a bottle of wine and a note thanking me for highlighting the issue. They knew I was not trying to score points. They accepted I was just protecting the company.

This was achieved without the company ever feeling the need to broadcast to the workforce and boast that they really cared about their staff. They did not need to issue numerous letters telling us how transparent the management was. We were never inundated with yet another survey asking us how the airline could be improved. All the crew knew it was an excellent airline to work for, not because of some PR spin, but via what we saw and experienced every day.

Airtours had over the years morphed into MyTravel. MyTravel eventually merging with its once fierce rival, Thomas Cook.

MyTravel's holiday arm of the business had not performed well and was on the brink of collapsing, dragging the airline side of the business down with it. Thomas Cook saw this as an opportunity. With an individual named Manny Fontenla at the helm, they decided that the billion or so losses of My Travel would be useful to offset any future profits of Thomas Cook. They also realised that by merging with its rival, it would, at a stroke, remove from the high-street one of its biggest competitors.

The more seasoned and experienced managers saw this as a perfect time to retire, maybe sensing the impending change of corporate emphasis. To confirm and underline this changing of the guard, the board decided to parachute individuals in from outside.

It has now become commonplace for airlines to pronounce they have policies in place that promote communication and address protections to anyone who expresses concerns relating to safety or illegalities. Since I had joined Airtours, there had never been a need for these grand statements. The pilot community took as a given that if anyone ever did raise concerns, then they could do so in full confidence that this would not have a detrimental effect on their future career. Virtually overnight, the entire culture and atmosphere changed with GCSE Business study buzzwords beginning to appear such as, 'just culture' and 'speaking out policies'. For anyone who had an experience of the real world, most of the pilots saw this for what it was. Corporate virtue signalling. Many of us were therefore not surprised when there was a rapid increase in the number of pilots suspended for extremely minor, non-safety related, transgressions. Whereas previously these minor transgressions would have resulted, at the most, with tea and biscuits and a one-sided conversation with the chief pilot, they now warranted suspension from work.

It appeared someone, somewhere, was trying to make a name for himself. That 'someone' was the new boy, chief pilot Paul Hutchings. For those of us who had 'been around the block', it was easy to see why this was happening. The external candidate who was now the chief pilot appeared to be using this period to flex his corporate muscles and display to his betters that he was not afraid to show the pilot workforce who was boss. Not that it would affect me, of course. Since I began this particular career path in 1989, at no time had I ever had a single disciplinary issue. Don't get me wrong, I had in the past had my decisions questioned, but one advantage of sticking to 'Standard Operating Procedures' (SOP's) is that if your decision was challenged, then you could refer directly to the reason why you made the decision you did.

Talking of decisions, I am convinced one of the main reasons I had avoided confrontation with any of my employers was due to a piece of advice I was given by a Saab captain some years previously. His name was John Smith. John was known to be a straight-talking, intimidating individual. For some reason though, he had taken me under his wing. It was one of my last flights as a first officer on the Saab when he divulged this pearl of wisdom.

'You know what the prime job of a captain is, don't you?'

'Don't prang it,' I replied.

'Don't make a decision until you have given them options...they can't touch you then.'

John went on to explain that many who sat in the office, be it crewing, operations or flight managers, had only one concern. They didn't want to be responsible when the wheels came off the operation. Their interest was not about the pilots or passengers. It was about making sure they would not be held to account for any costly disruption. But as John eloquently highlighted:

'Their arse is not on the line travelling at 250 knots mate...remember that, and you won't go far wrong.'

This advice had stood me in good stead several years previously during my command training.

There were early indications of where the airline was heading with Hutchings as chief pilot.

I was down route in Goa, India. My fellow crew member Paul was an exceedingly amiable chap. He was at the time also my pilot manager. After the usual amazing curry, and no doubt lubricated after several Kingfishers, Paul announced:

'You are a fucking nightmare to manage.'

' Sorry?'

'You. You are a fucking nightmare to manage.'

He then went on to explain that on several occasions since Hutchings had become chief pilot, he had been contacted by the ever-expanding HR department and told to call me in for a meeting over some perceived *misdemeanour*. He explained that every time this happened, he would have to tell the HR department that the perceived misdemeanour was due to my complying with what was set out within the company manuals. The manager said he felt like a piggy in the middle in that the HR department was being pressured to raise the issue from senior management. He was also fully aware that if he did call me in for a meeting, I would just tell him to tell HR to read the Thomas Cook operations manual. We both agreed that this situation rarely, if ever happened, before the arrival of the new chief pilot.

As he took another swig of the Kingfisher, he told me:

'I would have to tell them you were technically right.'

'No not technically right, completely right,' was my response.

'See you are doing it again: you never give an inch, that's exactly what I mean—a fucking nightmare to manage.'

The line pilots had an ally, at least while this guy, an experienced pilot himself, was our immediate manager. He was the fire break between normality, HR, and Hutchings' world.

It was around this period that I got the first real sense of the future corporate direction of Thomas Cook. Previously there was a definitive feeling through the entire structure of the airline that we were all working for the common good. The pilot community could sense that this was no longer going to be the case. The company's culture was shifting. It appeared it was going to be based on an 'us and them' philosophy. The 'us' in the equation being the pilots, cabin crew, office staff and Thomas Cook passengers. The 'them' being the senior management and directors who, at least from my perspective, seemed more interested in corporate bonuses than overseeing a harmonious working environment which placed the customers at the heart of the operation. The following are examples of the above culture.

Every year there would be a junket to a Thomas Cook destination. Here, several cabin crew, a sprinkling of pilots and Thomas Cook high-street staff would spend a couple of days in a resort while being indoctrinated by Manny and his cohorts. They were told how important they were, collectively and individually, to the Thomas Cook brand. Manny would recount that the passenger experience was defined by the interaction of the cabin crew with the passengers. I had been told that every year, on cue, Manny, with his shirt unbuttoned, would become emotional about how proud he was to be working with such a fantastic group of employees, dabbing his eyes as he teared up. The cabin crew and high-street staff lapped it up to the extent they would be standing on chairs and cheering on this bucket and spade Messiah. No doubt the free-flowing beer and wine helped define the atmosphere. Such was the loyalty that this little act instilled within the inebriated *fully inclusive* disciples that anyone who dared question him was seen as a *half board* Judas.

Manny was so proud of the employees that he set up the PROUD awards. They were established to recognise individuals and sections of the company that had done most to contribute to the success of Thomas Cook that year. One particular year the cabin crew were quite rightly recognised as being the most significant profit centre of the airline due to the amount of duty-free they had collectively sold. As a result, Manny and the directors awarded themselves massive financial bonuses. The directors got their millions, the cabin crew got:

A monetary bonus?

A holiday voucher?

No. For being the PROUD award winners, the cabin crew were presented with a lanyard to attach their security pass to. The lanyards were decorated with text announcing that the crew were indeed, collectively, that year's PROUD award winners!

The reality of how a number of the Thomas Cook senior management and directors viewed their own self-importance over those that were vital for the success of the company, namely the passengers, was laid bare one very early morning at Manchester airport.

It was still dark when I landed the Airbus A330 back at Manchester after flying through the night the 5000 miles from Cancun, Mexico. I had three hundred and sixty-five passengers on board. There had been issues with the IFE (in-flight entertainment system) not working, the cabin lights had come on unprompted waking the slumbering passengers throughout the night, and a couple of the toilets had ceased to flush meaning queues at those that were still operating. These and similar other issues were becoming regular occurrences. The cabin crew, again as per usual, were taking the brunt of the passengers' frustrations.

There again, they could always look at their PROUD lanyard and get a little warm glow inside.

Eventually, to the relief of everyone on board, we landed at Manchester and began to taxi the last few metres to our final destination, which was gate 32. Gate 32 has an air-bridge which allows passengers and crew to walk directly off the aircraft and straight into the terminal. From shutting down the engines to the last of the passengers disembarking would take around ten to fifteen minutes. As I was about to make the 90° final turn to align the A330 with the stand guidance system, ground control contacted us. We were now instructed to make our way to a remote stand. This stand had no services and no air-bridge. It was at the back of beyond at the airport. I brought the aircraft to a complete stop to challenge this nonsensical clearance. I wanted to know why, literally, at the very last moment, were my passengers and crew going to be severely inconvenienced. If we did have to park on a remote stand, due to it all being very last-minute, experience would dictate it would take anywhere between 30 minutes to an hour to arrange for steps and the several coaches that would be required, to arrive. I was told by the ground controller that my clearance to taxi on to stand 32 had been cancelled at the request of Thomas Cook operations themselves. He suggested I should contact the company directly to find out why my passengers were going to have to wait up to an hour before they could even begin to disembark.

As soon as I had shut down the aircraft engines and called the handling agent to arrange the steps and coaches, I contacted Thomas Cook operations to find out exactly what was going on. They informed me that the reason for the late change was due to a VIP arriving on a Thomas Cook 757. It transpired the VIP concerned was Manny and a handful of his merry band of company directors returning from this year's sermon on the piss junket. At least my suspicions were now appearing to be confirmed about the culture at Thomas Cook. Despite what was said about the customer coming first, the reality appeared somewhat different. The last of the Cancun passengers disembarked the aircraft seventy minutes after we had landed. Manny the Messiah did not walk on water, but he did walk on an air-bridge, unlike his passengers.

There had been a sea change of emphasis within the culture of the airline. This was not going unnoticed. So what was behind this sudden switch in focus? Was it just coincidence that this appeared to coincide with the rapid promotion of Paul Hutchings from chief pilot to the director of flight operations (DFO) for Thomas Cook airline?

With the new incumbent now in post, it was apparent that this individual was now intent on building a team around himself. In itself, this was completely natural. What was not understandable was that some individuals were being placed in roles, not based on ability or experience but, it appeared, based on a willingness to be prepared to act in a subservient manner to the vagaries of a director's whims. In other words, yes men and women.

The aforementioned pilot managers positions being a perfect example. Hutchings had decided to remove all the current pilot managers. At a stroke, out went the buffer, the voice of reason, between the pilots and senior management. Until this point, pilot managers were experienced pilots who understood the rules and regulations by which we were constrained. Hutchings' master plan was to replace these knowledgeable individuals with ex-cabin crew or HR officers. So why exactly would the new DFO willingly put individuals with zero knowledge or experience with regards to the stresses and strains and requirements that pilots face daily into such a position?

Previous pilot managers had rigorously protected the line pilots when senior management or HR overstepped their remit.

The new ex-cabin crew, now pilot managers, appeared only too willing to blindly accept as gospel what they were handed by HR and senior management. Was this due to lack of confidence, knowledge, or was it downright incompetence? Most of my colleagues suspected it was a misplaced sense of loyalty to those who had promoted them, quite unfairly to all concerned, way beyond their apparent ability, background or knowledge.

But of course, maybe that was the whole point. Perhaps Hutchings needed individuals who would toe the company line without asking too many questions. Was it a case of Hutchings feeling that, in order to secure his own position, he needed lower managers prepared to do his bidding without letting facts get in the way when investigating any alleged misdemeanour?

The immediate result was that there was a rapid increase in experienced airline pilots who, for the most part, had no previous disciplinary issues, now suddenly being suspended and given written warnings for the most minor of infractions. This culture could quite easily lead to a substantial reduction in in-flight safety reports. A workforce who would rather conceal any incidents for fear of disproportionate disciplinary action taken against them is no longer a safety centric operation. Yes, the company had installed the speaking out policy, but why would anyone speak out if it caused a detrimental effect on their future career. I was not alone in suspecting this was precisely the threat that Hutchings was banking on to prevent dissenting voices.

Since the beginning of my flying career, I had always complied with company rules and legislation. It had always stood me in good stead. I was bulletproof. It meant that while others were being dragged in by the new pilot managers for disciplinary meetings, I was just getting on with what I was paid to do, moving people from one side of the world to another safely and as comfortably as I could. That said, within weeks of Hutchings becoming DFO, I was hearing reports that more time was spent by my colleagues looking over their shoulders so at least they could see where in the back they were about to be stabbed.

Hutchings had appeared to have surrounded himself with yes men and women. So had Manny.

When flying, either pilot can call 'go-around' and abort the landing. The absolute worst trait in a pilot would be a 'yes man' mentality. No one appeared to call a business plan 'go-around' when presented with the equivalent of a 747 on the corporate runway ahead. The result was, for the first time in its history, Thomas Cook was beginning to see the cracks that would quickly become, a yawning cavern in its finances.

Anybody who has ever seen, or heard or even gone as far as using this thing called the internet knows one thing: business is moving there. The number of empty premises on the high street is testament to that. It's cheaper, easier, and more versatile than having a bricks and mortar presence. So what was Manny's, surrounded by his yes-men, next idea to move Thomas Cook forward? It was to buy, from the Co-Op, all its high-street shops, paying top price for them into the bargain. His directors and managers were fighting over themselves to pat Manny on the back over his brilliant idea that would buck the trend of every other company who were now concentrating on their online business.

I was not alone in suspecting that senior directors and managers at Thomas Cook saw their position as a short-term opportunity to rape the company with inflated bonuses. It appeared passengers and crew were an annoying necessity required only to collect those bonuses. We were nothing but stepping stones!

So what happens if the criteria set to collect those bonuses were not reached? One such measure was the share price. The city and investors were not impressed, and the shares had taken a substantial hit in response to Manny's decision to buy the Co-Op high-street shops.

Remember though, Manny could turn the water of a terrible business idea into the wine of a few million in directors' bonuses. Was this divine intervention? No, he just claimed they had set the share price bar that would trigger the director's bonus too high in the first place, so he lowered the target, after the effect.

So what happens to those not beholden who have not fallen under the spell of Manny, Hutchings and the merry band of carpet baggers? Did I mention I had never had any disciplinary issues in my entire flying career before?

# CHAPTER NINE

## *Lavish With Ourselves , Lean With The Customer*

*"Pepsi will be loaded on this route, not Coca- Cola, as this is made by a Jewish owned company."*

UK Airline BMI

The day started like any other. A drive to work listening to B.F.G.'s tribute band, Nine Inch Nails, and wondering in what state the Airbus 330 I would be presented with that morning would be. Over the previous weeks, although not safety-related, issues that would impact directly on passenger comfort and enjoyment seemed to have been put on the back burner with regards to fixing them. It would result, as it always did, with my cabin crew colleagues having to face yet another barrage of angry passengers not getting what they paid for.

With Nine Inch Nails' Trent Reznor's whingeing still in my ears, as per usual, I arrived at the crew room fifteen or so minutes before report time. Today's flight was to Orlando, Florida. Arriving early before a duty was not due to some altruistic desire to show the management that I was willing to spend my own time preparing for the flight. No, the reason was that the staff car park was such a distance away from the crew room that it necessitated catching a staff bus that ran during the day every twelve to fifteen minutes. At night it only ran every thirty minutes or so. That realistically meant, to ensure I would not be late for a report, I would have to arrive at the staff car park around 30 minutes before report. This extra time was not included in any duty time calculation by either the company or the CAA.

Things were looking up. The first officer that day was Andy, a fellow drummer and an all-round top guy. Andy had already printed out the paperwork and was going through it with a fine-tooth comb. Having had a scan over the weathers, I asked my usual question 'which way do you want to fly it?'

'With my hands,' came the reply. I should have known.

Very unusually, given the past few weeks' experiences, there appeared to be no snags, not even minor ones on the Airbus 330 we were due to fly in an hour or so. We both concluded this was due to the snags not being reported or, alternatively, the company was finally taking the full customer experience into account. As myself and Andy were deciding on a fuel figure, one of the new, shiny, Hutchings-appointed pilot managers wandered over. Her name, Jo Burke. She was previously cabin crew and had never flown commercially as a pilot. 'Just to let you know, you have two directors on board today.'

Maybe it was the look on both mine and Andy's face that prevented Burke from suggesting we hurry up and get down to the aircraft early. Now it made sense. This was why there were no snags on today's aircraft.

As we walked along the air-bridge connected to the Airbus 330, we noticed that the floor was wet. It had not been raining that morning, so this was unusual.

Once we had both dropped our flight bags into the cockpit, Andy began busying himself entering the route into the computer from the printed flight plan. I did the walk around checking we still had both wings attached and no one had run off with the wheels. As I came back into the rear galley of the Airbus to say hello to the crew, one said: 'just look at them'.

She was referring to the several cabin crew managers who were busy interrupting the crew trying to prepare the aircraft for the passengers. It was a shame the same cabin crew managers did not have the same dedication to duty to appear on the aircraft when Manny and his gang had caused us to park on a remote stand. Of course, we didn't have a couple of company directors on board that morning. The stench of sycophancy was hanging heavy in the air.

As I walked towards the front of the aircraft, I eventually arrived at the front row of passenger seats. The base manager was there, flanked on either side by two men with vacuum cleaners. Again, it was unusual to see this base manager on a turnaround, even more surprising to see her directing cleaning operations. The base manager was forensic as to where she wanted these vacuum cleaners aimed. As the two cleaners got to work, she started to cross the seat belts on the two seats in front of her. I introduced myself.

'I understand we have two directors on board today?'

'Yes we do,' the base manager replied. I enquired where they were sitting.

'Here and here.' She pointed to the two seats where the carpet had been vacuumed to within an inch of its life. It also coincided with the seats that the belts were crossed.

'Why is it that we don't make the same effort for the passengers as we do for the directors?' I went on to explain that the reason I asked that question was that it was the first time I had flown the 330 in a long time that had virtually zero cabin defects. Whoosh, I had lit the blue touch paper. The base manager frantically attempted to convince me that every turnaround was given the same treatment.

There was no point in continuing this conversation. I made my way towards the cockpit. There were now two engineers sitting in the flight deck. I explained what the base manager had told me.

Both engineers began to laugh. 'You should've been down here 15 minutes ago. They called someone specially to mop the air bridge and then spent 10 minutes polishing the step into the aircraft.' Well, that explained the wet air bridge.

I called one of the cabin crew, to ask the base manager to come back into the cockpit.

Once she arrived I asked her directly 'You told me extra was not being done because we had directors on board, but now I find out, not only are you crossing seatbelts and directing cleaning operations but you also arranged for the air bridge to be mopped and the step into the aircraft polished.'

After her initial denial, she relented 'Ok, well we want it to look nice for them. We would do the same for you'. Somehow via a superhuman effort, I was able not only to keep a straight face but also prevent my eyes from rolling.

Once we had arrived in Orlando, I had arranged to meet Andy in the bar for a drink before we headed out for a steak. With a coke in hand, I had a few minutes before I needed to leave the hotel room and head down to meet him. I had a quick look at the company website forum. A crew member, a couple of hours previously, had posted that the state of the Airbus 330 fleet was appalling. They highlighted it was an embarrassment to repeatedly have to face the valid criticism from the passengers. He suggested that it would be a good idea for the directors of the company to fly on the aircraft just to see how bad the situation now was with the fleet. I posted a concise reply. I stated that there was not much point in the directors getting on a flight because when they did the lower ranking managers would bring out the *'premium wash and go service'* in an attempt to conceal just how bad the situation had now become. I hit reply and went to meet Andy in the bar.

The two directors were already sitting at the bar and already seemed to be worse for wear. They were also extremely loud; however, they did look very popular. Was this because they were buying drinks for everyone? Was this on the company credit card perchance? Andy and I looked at each other. We did not need to say a word. The last thing we wanted was a night of false fun. We turned around and headed straight out for the steak. It was excellent. Vitos on International Drive, I heartily recommend it.

The following week I had a change to my roster. That Friday evening I *was* due to drive to Glasgow for a duty the next morning. The change to my roster now reflected a meeting with the new pilot manager, Jo Burke.

I rang Burke up and asked what the meeting was about. At first, she told me she was not at liberty to spill the beans. Surprisingly, she seemed a little taken aback when I told her I would not be attending any meeting until I knew about the details. Eventually, she told me it was to do with something that I had posted on the company forum. I asked her directly:

'Have I put someone's nose out of joint?'

After a few umms and ahhs, 'yes' was Burke's reply.

'Are you caught in the middle of this?'

Again she eventually answered 'yes'.

Burke then informed me that 'it's really important that we meet to get to the bottom of what went on'.

I explained we didn't need a face to face meeting, it was merely a case of lower management trying to conceal from directors just how bad things were, and that the base manager had got upset when I called her out over it. It was as simple as that.

'Yes but Paul says I still need you in for that meeting.'

So why exactly would Hutchings, the director of flight operations with enough issues on his plate, want to get involved when one of his senior captains had raised the point that passengers appeared to be short-changed compared to the directors? I explained to Burke that I would not be attending a meeting at three pm, then having to drive home, only to then drive past the airport that evening on the way up to

Glasgow for a bullet Orlando.

'When do you get back?' Burke asked.

I explained that, after having been up all night flying from Orlando to Glasgow, I would then position back to Manchester as a passenger after having been up for around 19 hours. Despite this, Burke was insistent that the meeting had to take place as soon as I arrived in Manchester. On arrival, she was there waiting for me.

She opened the meeting by explaining that Hutchings had demanded that I write a letter to the base manager to 'apologise for my behaviour' in confronting the base manager.

If it hadn't been clear beforehand, it was now crystal clear, at least from my perspective. The issue wasn't how can we turn this company around so that the passengers benefit. It was, *don't you dare question the motives of, or embarrass the directors or anyone attempting to conceal the reality of just how bad things are getting.*

I explained to Burke, all I had done was raise the issue that if so much effort can be made to ensure the aircraft is fully serviceable and clean when we have directors on board, why can't we do that every time.

Burke then pulled out the company's well used and utterly predictable fall-back position. I had been pre-warned that this was used whenever the managers hit a dead end.

'It was the way you talked to her.'

I suggested, in that case, it would be a good idea to interview Andy, the two engineers and the cabin crew, all of whom witnessed the discussion. For some reason, Burke did not want to do that.

Suddenly Burke announced 'I also have this letter from Maggie Kennedy'. Kennedy was Director for cabin crew services.

It appeared Kennedy was now also putting the boot into my character. She described how rude I had been to the base manager and how I was ruining the teamwork that they were building.

I inquired, 'Does Kennedy mention anywhere that she has never met me?'

By this time, I was starting to feel sorry for the ex-cabin crew, totally out of her depth, pilot manager. She was, after all, as she had previously admitted, only doing what she was told to do.

'Well, Paul still wants you to write a letter of apology.'

'Does he now? Well, you can tell him I will send a letter, but it won't be one containing an apology.' With that, I got up and left.

Well, here I was, the first time since I started flying in 1989, having my integrity questioned. I sensed, given that I had not given an inch, the next few months or so might get very interesting indeed.

I did indeed send a letter to the new DFO. I suggested that instead of attempting to silence criticism, it would be beneficial to all if my concerns were listened to and dealt with appropriately.

A couple of weeks later I was once again up in Glasgow. I was sat in the airport hotel bar chatting to a local captain. He had heard about my little set-to with the management.

'You know those two have form, don't you?'

He went on to explain that he had flown the same two directors back from Orlando a month or so previously.

He described what had happened on his flight and that he was so disgusted that he also posted what he had seen on the internal company forum.

Almost immediately, his posting was taken down. A few minutes later, he had a call from one of the new pilot managers, instructing him to apologise to the two directors. He refused. What happened next highlighted precisely the culture that had taken hold.

The captain's wife, who also worked for Thomas Cook as cabin crew, was called in for a meeting. She was informed that her husband had refused to apologise to a director, and therefore, *she* was to issue an apology on his behalf. She was advised that should she refuse, then things would become 'difficult' for her at work.

'She didn't, did she?' I asked him.

The look on his face told me that she had.

This was all starting to sound familiar. I had seen this mentality before.

The veil had been initially lifted on the inner workings of the aviation industry when I became involved in a case on behalf of my wife Lisa when she worked for the British Airline BMI.

BMI had agreed to operate flights to Jeddah in Saudi Arabia, as British Airways had abandoned the route. BA cited economic grounds for their decision. Press reports had highlighted the reason it had become no longer economic was due to the crew objecting to night-stopping in Saudi. There had been several terrorist strikes on westerners and western compounds and the BA crew's determination, quite understandably, was that it was no longer safe to stay in Saudi. This constrained BA to night-stopping its cabin crew and pilots in Cyprus or Kuwait, resulting in extra expense.

BA could not have dropped the route at a worse time. Again, the press reports suggested that the Saudis considered this act a severe slap in the face from the UK. At the same time Tony Blair, the Prime Minister, was being urged by the Saudi Government to drop a corruption case against BAE, the UK arms and aircraft manufacturer, relating to the al-Yamamah project. The case involved an alleged slush fund and a high-ranking Saudi prince. In no uncertain terms, Blair was intimidated by the Saudi Government to intervene and halt the case being constructed by the Serious Fraud Office. Blair cooperated with the oil-rich state, and the case was quietly kicked into the long grass.

Where BA feared to tread, BMI said 'hold my beer'. BMI, apparently, had little issue with their crew staying in Saudi, even during the terrorist threat. BMI had also informed their female staff, via a memo, that they must remove any outward signs of their religion, such as a crucifix, while at the same time demanding that they wear the Islamic abaya. Furthermore, the female crew were instructed that they must walk behind their male colleagues, irrespective of rank.

The Chairman of BMI, Michael Bishop, was subsequently made a Lord by the UK Government. His title now is, Baron Glendonbrook.

This was where I first became involved in dealing with the dark underbelly of segments within the airline industry. The experience laid bare the willingness of some senior managers to lie, collude and bully employees in the pursuit of maximising profit.

Up to this point in my career, I had never witnessed anything quite like it. It was a side of the business that, understandably, the airlines, unions and governments wish to keep hidden from the media, public, and staff.

Lisa, my now wife, refused to operate to Saudi under those restrictions and was summarily dismissed. Her union, Unite, declined to intervene. Lisa consequently lost her legal action against BMI. The reason she lost, to this day, remains unfathomable. It was at this point I personally became fully involved in the situation and swiftly discovered that the only language airline managers appeared to understand was the one that directly affected the bottom line and therefore bonus.

It was a case that ultimately involved the British Government at Cabinet Level, the European Parliament, Israeli Government, Buckingham Palace, Lufthansa and, as a result, Michael Bishop's bank balance. The full story became such a twisting political affair that the original screenwriter of 'Homeland' and 'The Spy', Gideon Raff contacted me, to provide the background to the story. I told him he would need to wait for the book.

Suffice to say, where commercial interests collided with integrity and morality, at least from my perspective, BMI had chosen commercial interests. One clear example of the lengths that the airline would go to was made apparent in a memo conveyed to BMI training staff. The British airline was not afraid to pander to one particular religion and to denigrate another in the pursuit of profit. The memo stated:

*"Pepsi will be loaded on this route, not Coca- Cola as this is made by a Jewish owned company."*

That one announcement highlighted, to my mind, the cultural illiteracy and lack of intelligence within the corridors of an airline operating into the most politically-sensitive regions in the world. It appeared BMI were desperate to pander to the predominantly Muslim passengers and were willing to fabricate an anti-Semitic reason to highlight why they only stocked Pepsi. It appeared the crew were expected to trust this incompetent BMI management with their lives. All, apart from two cabin crew, did.

The details of the case were subsequently described as explosive by one of the U.K.'s leading barristers. I will, however, save this particular story for another time and another volume.

The experiences and lessons learned during this debacle, however, provided the perfect grounding for what was to follow only a few short years later.

I'm still waiting for a reply from my letter to Hutchings.

# CHAPTER TEN

## *Jam Tomorrow*

*"You're fucked, I'm fucked, we're all fucked."*

Paul Hutchings DFO Thomas Cook

Unsurprisingly, things hadn't been going well for quite some time at Thomas Cook. Appalling management decisions, undeserved director bonuses, and a culture that was intent on suppressing any dissension was rife. Frankly, I had never seen anything like it. Neither had the City. The share price had tanked, and Thomas Cook had been dumped out of the FTSE 100.

Like many, I was spending nearly £100 a month on membership of the pilot's union, BALPA. This was for legal assistance as insurance should I come into legal conflict with the authorities or for any employment issues. Since Hutchings' arrival, I considered the likelihood of that happening had soared from a rank outside possibility to an odds-on racing certainty.

BALPA is, in effect, structured in two tiers. Firstly, there is the head office at London Heathrow from where the full-time staff operate. Each airline then has a company council. The company council, or 'CC, is made up of pilots who fly for that particular airline. The role of the CC is to deal with pay negotiations and other issues that affect the pilot workforce as a whole at that specific company. They also provide support should an individual pilot come into difficulty with the airline in the form of either disciplinary meetings or indeed if the pilot wishes to register a grievance.

BALPA head office deal with the more significant issues. Those concerns that affect the entire UK pilot community. They are also there to provide legal advice and support to an individual pilot if any employment issue escalated to a level that required the involvement of a solicitor or barrister.

During this period, the chairman of the Thomas Cook BALPA CC was an individual by the name of Roger Scadeng. Up to this point, I had zero dealings with Scadeng, or with the CC for that matter.

The BALPA CC, under the leadership of Scadeng, had previously been involved in pay discussions. Things had got so bad under Manny that, in lieu of a pay rise, the company agreed to issue shares to the pilot community. It was a similar deal to the bonus structure that was enjoyed by the directors, but, somehow, just that little bit *different*. The feeling on the line was that this was a 'jam tomorrow' promise. The Thomas Cook share price at the time was virtually worthless and the likelihood of us ever having shares issued to 'share in the success,' was so far off on the distant horizon it was barely worth thinking about. Manny had led the company down the wrong path, yet he was still collecting a healthy few million in bonuses, as had several who had agreed with him. My personal view, this guy was a chancer. The reality was, the pilots endured another year of an effective reduction in salary due to inflation.

Such was the ever-growing distrust of Manny by the City, in August 2011 his previous enablers, the board, kicked him out. The wreckage of his tenure however remained. By July 2012, he had been replaced with a new CEO by the name of Harriet Green. The City loved our Harriet, and almost immediately, the share price and all of the associated fundamentals began to improve. In a BBC radio interview, Green described herself as a 'Landa' (sic). When asked to explain what Landa meant, Green replied that sometimes she had to roar like a lion, but she also gave cuddles like a panda. Some of you may now be confused and wondering why I am now talking about an infant school teaching assistant. No, I'm still referring to the new CEO of Thomas Cook.

Although the company share price was now slowly beginning to rise, it still looked like the pilots were going to take a hit. The airline, with Hutchings in charge, decreed that radical changes were needed. It was then, in late 2012, that Hutchings announced that compared to the likes of easyJet, our costs were too high. He also declared that eleven of the aircraft that the company leased were up for renewal. To persuade Landa that the current in-house airline was a far better bet than relying on third-party companies, Hutchings stated pilots, collectively, would have to face a significant salary cut. As if this were not bad enough, we would also suffer a substantial reduction in pension contributions, numerous redundancies and many demotions from captain to first officer. On that basis, he announced that a consultation period had begun with BALPA. At this stage, the company recommendations would be analysed, debated and then considered during regular consultation meetings. Roger Scadeng would head the BALPA delegation. Hutchings himself would represent Thomas Cook along with another senior manager.

It would conclude with the CC recommending to the Thomas Cook BALPA membership to vote and accept these changes or, a proposal to vote to reject the scheme. Once the CC had made their recommendation, it would then be up to the pilot membership to vote one way or the other. The trouble Hutchings faced was, like the little boy who cried wolf, we had heard this kind of threat so many times before that, for the majority of us, it now fell on deaf ears. Few believed him anymore. The belief was that the intention was to simply get costs down so that directors could justify yet another bonus and to engineer a situation where Thomas Cook could achieve what they had wanted for many years. The pilots to work part-time.

A colleague later informed me of a presentation he attended with several other pilots around this period. Chairing the meeting was Hutchings. It was during this gathering that Hutchings, aided by a PowerPoint presentation, attempted to convince the assembled crowd that Thomas Cook pilots were significantly less efficient compared to the pilots at Ryanair and easyJet. The thrust of Hutchings' argument was that every year, easyJet and Ryanair achieved far more flying sectors from their pilot workforce than Thomas Cook. He appeared blissfully oblivious to the somewhat undeniable fact that Ryanair and easyJet were exclusively short-haul operators. Thomas Cook pilots also flew a substantial long haul programme. He was attempting to compare *apples to oranges.*

My colleague then asked Hutchings a reasonable question. How would it be possible for a Thomas Cook long haul pilot to fly as many sectors as a short-haul operator? Apparently, this challenge came as somewhat of a shock to Hutchings. That one, simple question had just dismantled his slide show presentation. Immediately he began to rant. His entire ruse blown. According to my colleague Hutchins response was;

'Look, if you don't fucking vote for it, you're fucked, I'm fucked, we're all fucked!'

It looked like Thomas Cook pilots would have to leave it to the CC, with Scadeng as the chairman, to represent our interests at the consultation stage. It was here the proposed significant reduction in salary, pension contributions, demotions and redundancies would be debated and argued.

I was then informed by a member of the CC that, subsequent to the consultations with Huchings, the union had held a Company Council meeting to debate the airlines proposals. Knowing exactly what the company were demanding, a straw poll was then taken.

According to this individual, the majority view of the CC, which included Scadeng, was that the proposals could in no way be seen as acceptable, or reasonable, and should be rejected. At that meeting, the consensus was that the BALPA CC recommendation to the pilot workforce would be to vote 'no' to the Thomas Cook proposal. However, the formal vote on what the CC would actually recommend to its members was to take place the following week.

The same CC member then informed me that, over the intervening weekend, Scadeng was allegedly seen in a restaurant with Hutchings. It subsequently became known to the workforce as the 'curry meeting'. Irrespective of whether this meeting did or did not take place, by the Monday when it came to the actual vote, Scadeng had mysteriously dropped his opposition to the proposals and, as chairman of the BALPA CC, now voted that they should recommend the plan to the workforce. Three of the CC members immediately resigned. Was this due to Scadeng's sudden, '*Road to Damascus*', conversion? I was subsequently given exactly the same account by another, now ex-member of the CC. True or not, this was what two of the BALPA CC informed me.

As is usually the way in these matters, the pilot membership takes great heed of the recommendations of the CC, and this was the case in this particular situation. Of course, it only came to the wider pilot community's knowledge that Scadeng had, allegedly, shared a naan bread with Hutchings *after* the workforce had voted.

Within days I received a letter from Burke. She informed me I was being kicked out of the Manchester base, outside of seniority, and outside of any agreement, while less senior pilots were remaining at Manchester. What the hell was going on?

Having discussed the events with a few more colleagues, I decided to give Scadeng a ring and talk to him for the very first time. I told him it was about the enforced base move.

'Everything OK?' Lisa asked as I hung up the phone.

'There is something extremely fishy going on.' I explained I had asked Scadeng the following direct and closed question:

'Was it, at any time, agreed between BALPA and Thomas Cook that pilots could be displaced from their home base out of seniority order?' Scadeng had refused to give a straight answer. He stated that everything that had been agreed on was in the proposal document that the pilots had voted on.

Having spent most of my twenties poring over complicated and detailed music label contracts, it was a five-minute job to give the BALPA agreement with Thomas Cook the once over. No, no mention of seniority being ignored. No phrase could be misinterpreted, misread, or misunderstood. It was cast iron. What Hutchings wanted to do was outside of any agreement. Therefore, as someone who represented the pilots' interests, why would Scadeng not answer a direct question?

During my first drive to the new base, I considered if I should just shrug my shoulders and accept this diktat. That would be the easiest thing to do. Everyone else seemed to be doing so. I then considered everything I had to go through to become an airline captain, the years of no money, no car, a massive amount of debt around my neck and having to travel to the other side of the world to get a flying job. If I raised a grievance over the issue, I suspected Hutchings and his gang would round on me like a bunch of hyenas. But this was the thin edge of the wedge, and I was not going to let it ride.

I contacted a long time friend who I have known for forty years. We first met when he had sent both BJ Williams and myself off in a football match in 1979. However, we quickly got over that rocky start to our friendship. Not only was my friend now a Thomas Cook pilot, but he was also a member of the BALPA CC, at least, he had been. He had attended the consultation meetings between the company and the union. He was one of those who had resigned in disgust at the CC voting for the pay cuts and redundancies after the alleged curry meeting. I asked him if, at any point, had this forced base move, out of seniority order, ever been agreed by BALPA. His answer was unequivocal.

'There never was a fucking agreement to do that.'

He explained that Hutchings had tried to claim it had been agreed at a previous meeting. Not only did he personally correct Hutchings, according to my referee friend, even one of Hutchings own senior managers stepped in to correct Hutchings' attempts to steamroller the issue. This was excellent news for my case. A letter from someone who had witnessed what was said first hand would shoot Thomas Cook's claim that my and others enforced moves, had been agreed with the union to pieces.

'Great, can I ask you to put it into writing for me that you witnessed that there was no agreement.'

'Errr no I won't be able to do that.'

I have to say I was a little taken aback by my friends' response.

'Hutchings is a nasty bastard, and if I did, he would come after me.'

The ex BALPA CC member went on to explain that because he had resigned from the company council, he no longer enjoyed the same employment protection as when he was a BALPA rep. He made it clear he had no doubt Hutchings would make his continued employment with the company untenable if he provided me with written confirmation that these forced moves had not been agreed.

Things were becoming more apparent by the minute.

I put in a grievance.

It was noticeable that the grievance meeting was not heard by any senior management. Instead, it was chaired by a 'rostering manager' named Amanda Richardson. I realised fairly quickly. I was getting nowhere using facts with Amanda.

'Would you agree that for there to be an agreement on any issue, both sides must agree that there is an agreement?' I asked.

'Sorry, what do you mean?' was her confused reply.

'Thomas Cook has displaced a significant number of senior captains out of base out of seniority order and claimed BALPA agreed to this in the meeting.'

'And?'

'Well, I have spoken to two individuals who were present at that meeting; one of them told me categorically that although Hutchings proposed this, it was rejected by BALPA at that meeting.'

I noticed Richardson squirming in her seat.

I then produced an email sent by Scadeng and read out exactly what he had written. "There was no specific agreement to alter any pre-existing agreements between BALPA and the company, other than that which was balloted on and accepted by the members."

'So?' was Amanda's curt reply.

This was like trying to reason with a pubescent teenager.

I produced the balloted document.

'Show me, anywhere in that document, that Thomas Cook can

remove captains out of seniority from their home base.'

Amanda's considered response, now it was checkmate?

'I'm not here to be put under a spotlight.'

I had all the facts. It appeared there never was an agreement between BALPA, other than in the fantasy world inhabited by Hutchings. So, why were BALPA not holding him to it?

My grievance was not upheld.

Here I was, kicked out of my Manchester base, and now flying out of Birmingham. The mileage difference between Birmingham airport and Manchester airport was less than half a mile. With that in mind, it would be reasonable to ask yourself, why did I raise a grievance if it affected me so little? On principle, ladies and gentlemen, on principle.

The principle was the airline had run roughshod over contracts and agreements and, more to the point, BALPA had allowed them to do it. This was the thin end of the wedge. In the music industry, if you let anyone rip you off, the word would get around pretty quickly that you were a soft touch. You would be eaten alive.

I was amazed that neither BALPA nor indeed a significant number of my pilot colleagues appeared to have worked this out. The one thing I could be sure of was that Hutchings would have felt significantly emboldened that his threats had achieved the desired effect. Not only that, as is the way in the world of business, those further up the slippery pole of management would likely be looking favourably upon his actions. Bonuses all round chaps!

There was one little ray of sunshine in all of this.

Since Landa had taken over the company, the share price had rallied significantly. Indeed, it had increased to such an extent that it would mean the pilots would be receiving their share bonus.

With the share price now at a level no one expected only a couple of years previously, based on the pay deal, the pilots could expect a windfall of around £50,000 to £60,000 worth of shares.

Well, actually no.

Hutchings was now attempting to convince the pilot workforce, and BALPA, that what we all understood about the share agreement was actually mistaken. No one was surprised.

One of the reasons proposed for not honouring the agreement? Too many of the company's shares would be held by too few people. At least that was one of their arguments.

It had been widely reported in the press that Landa had pocketed a golden hello when she joined Thomas Cook in 2012. Was it just coincidence the total amount of shares that the pilots were expecting, in round figures, equated to those awarded to Landa. So much for too many company shares being controlled by too few people. How does *one* sound?

Given the comments on the BALPA forums, it confirmed I was not alone in coming to the conclusion that it was a simple case of Thomas Cook attempting to renege on a pre-agreed deal.

Again, BALPA did not take the company to court and, with Scadeng still at the helm, the CC recommended to the pilot workforce that they agreed to a revised deal from Thomas Cook. We received about 10% of that which was due.

The regular conversation on the flight deck during this period was, why had the BALPA CC, appeared to have rolled over so quickly to Hutchings' demands? Pay cuts, demotions, relocations and now the share issue. I had smelled something fishy several months ago, and the stink was definitely getting stronger.

Within two months of this debacle, rumours started to circulate that one of the BALPA CC had applied to become the new Thomas Cook chief pilot. The BALPA CC, via the forum, tried to claim this was a wicked, evil rumour. Indeed, the vice-chairman of the CC stated:

"In more obvious terms. Why do people listen to rubbish when they ignore the truth?"

It wasn't rubbish, and it was the truth. By March, Scadeng was installed by Hutchings as the new Thomas Cook Airline's chief pilot. It said so much and answered so many of my questions.

# CHAPTER ELEVEN

## *Who's Witness Statement?*

*"It is an offence at common law to interfere with a witness by unlawful means, such as violence, bribery, threats or improper pressure."*

The Criminal Justice and Public Order Act 1994

It was November 2013 when the phone rang at home. It was a rare day off from the incessant short-haul and long-haul flights, it was a solicitor. He wanted to know the details of a flight I was in command of in August 2010. I had little inclination to discuss company business on a day off.

Over the last few months, as far as I had concluded, Hutchings had made it clear that neither he nor Thomas Cook could be trusted to abide by agreements. The evidence also suggested that BALPA would not hold the company's feet to the fire when they didn't. Had Scadeng, with accepting the chief pilots' position, finally made it clear where his priorities were all along? I simply joined the club. I made it explicit to the solicitor, this was company business, and he needed to contact the company and roster a meeting on a working day.

A few days later, according to my roster, a meeting had now been arranged at Birmingham Airport. In attendance, along with the lawyer, would be another of the new style, no experience, no idea of the rules and requirements of being a pilot, pilot managers. This particular individual had not even worked as cabin crew. She was from an HR background.

The question was, why had Hutchings considered this individual, along with the other ex-cabin crew, suitable to replace experienced pilots in the role of *pilot manager*? Were they more likely to do as they were told, conforming with missives from the flight operations management?

After my one hour and fifteen-minute drive, I arrived at Birmingham Airport. Compared to Manchester Airport it was luxury. The car park was a five-minute walk from the office. Well, it was if the hole in the fence was still there and you were prepared to pick your way through the mud, break branches and climb over wooden fences. Was there a better way to start your day than attempting a cross country obstacle course? There was always the walk to the monorail station if you didn't fancy communing with nature and could spare fifteen minutes.

Ten minutes after the meeting was supposed to have started, I was still the only one in the office. I rang the pilot manager. Naturally, for a meeting in Birmingham, she was still in East Midlands. She then explained, no it wasn't a meeting, it was to be a phone call. Not for the first time, or more importantly for this story, the last, the roster was a work of pure fiction. No apology, no nothing from the ex HR officer.

As I headed back to the car to complete the second half of a needless round trip of around one hundred and twenty miles, my phone rang. It was the solicitor.

He had a few questions. It was about something that happened over three years previously. The solicitor asked what I could remember about flying an Airbus A330, registration OY-VKG, on 20 August 2010. I asked him could he remember what he was doing on 20 August 2010. His answer was not without looking in his diary. I told him, considering I was currently holding a branch back to stop it springing back in my face, I did not currently have my diary to hand.

I reminded him it was over three years ago, so without him providing a hint or reminder, there was not much I could say. The solicitor mentioned it was a flight with another captain with the name of Garry. Amazingly, I did remember this flight. I had worked with Garry at Business Air, and I had only ever flown one trip with him. Unsurprisingly it was yet again to Orlando.

The solicitor asked what I remembered about taxiing out from the stand to the runway that day? What I did explicitly recall was that Garry, another captain, was sitting in the first officer's seat on the outbound leg. I also remembered that after we had started the second engine and it had stabilised, the pushback team and tug had disconnected and had driven away. It was at this point we experienced a failure of one of the flight control computers. I could not remember which one. This was not an uncommon issue as when the generator from the second engine came on line it occasionally caused a power spike.

The typical trick of turning it on and off usually fixed the problem. It was a computer, after all. That particular day, however, it did not clear the fault, and that was why I could accurately remember it. We requested to taxi back onto stand. We were then cleared to taxi the ten to twenty meters back from where we had just departed.

Once back on stand, an engineer had come on board and was able to clear the fault. Without further incident, we were on our way to Orlando. The solicitor thanked me for my time and that, I assumed, was that.

Five months after the conversation with the solicitor, late in the evening, I saw an email in my junk folder informing me that my signature was required on a witness statement by midday the next day. It was to be provided to a judge with regard to an 'incident' that had occurred on 20 August 2010.

I rang the solicitor the next morning to enquire what the hell was going on. He informed me that he had put together a witness statement based on our phone conversation and that he was here to protect me. He seemed somewhat put out when I reminded him that as Thomas Cook was paying him, I strongly suspected it was Thomas Cook he was protecting, not me. I also highlighted that at no time during our phone conversation in Birmingham did he inform me that what I was saying would form a witness statement. The solicitor had not even disclosed what the issue was all about. If he had told me that the conversation was to be used to 'construct' a witness statement, I would have refused to discuss the matter further. At least that is, until I had taken independent legal advice.

I flatly refused to sign the statement that was being metaphorically waved in my face. There were way too many inconsistencies and attributed comments that simply were not true. I did, however, say I would provide an account, but in my own words. I also explained I still had no idea what had been alleged. It was only then that I was sent the particulars of the claim.

I was told it was a simple case of a baggage handler trying his luck. He was attempting to sue all and sundry for an injury he claimed to have incurred. It was not just the airline he was gunning for. It was Manchester Airport, the airline and his employer - Service Air. His solicitor, probably a no-win-no-fee jockey, presumably had suggested he throw in the captain for a full set. According to this individual, the jet blast from the engines as we had taxied back onto stand had caused a baggage container to collide with him. If it did or it didn't, I have zero idea.

As I had promised, I wrote my own witness statement. I explained what I could remember. Fortunately, the Airbus 330 monitors the actions a pilot might make, even when still on the ground. Taxiing faster than the allowed 30 knots, not setting the flaps correctly or using excessive thrust while taxiing are all observed by the computers. If parameters are exceeded, the aircraft will notify the company directly. You could then expect a phone call from a member of the management team to explain yourself. No one from Thomas Cook had contacted me about this case in the three years since the alleged incident. I imagined the fact that Thomas Cook had not contacted me would indicate that no taxi parameters had been busted, including the use of excessive thrust. I considered it would be necessary for a judge to be aware of this fact. In my witness statement, I wrote:

"It is also important the court be aware I was only shown documents relating to the case on 16 April 2014, and at no point have I been provided with any details of the allegations by Thomas Cook management. I have not been contacted by Thomas Cook management and asked to provide details of the above incident until an email of 16 April 2014 asking me to contact the legal representative. Therefore, my recollection of events of an incident approximately four years ago are, understandably, rather distant."

I then went on to point out everything, both myself and the other captain did that day was entirely in accordance with Standard Operating Procedures (SOP's). I knew that because if either had done something that we shouldn't, the other pilot would have immediately called it out. The fact that we did everything *by the book* was also confirmed in the on-board monitoring system. If we hadn't, I would have had a one-sided conversation with a manager.

I then added:

"A normal taxi to take off would involve two engines to obviously ensure both are working correctly before takeoff, however, the company SOP is to turn one engine off after landing for taxi onto stand using only one engine. Therefore, it is company policy and acceptable to taxi with one engine. This, of course, requires significantly more thrust from the one operating engine rather than the power and attendant jet wash being halved for each engine when taxing on two." I then explained:

"Ultimately, the taxiways are for aircraft movements, and it is incumbent on all those working airside (including baggage handlers and engineers) to ensure that they keep a safe distance from a taxiing aircraft."

The solicitor was not happy. He informed me that they would not present my witness statement to the judge. The fact that I had highlighted Thomas Cook had not contacted me in nearly four years about the issue, according to the solicitor, would reflect badly upon the company. The other issue they had was my reference to taxiing with one engine as opposed to two.

My simplistic view was, if I was required to tell the truth, the whole truth and nothing but the truth, I was not going to put myself in the position of only telling the truth interpreted by Thomas Cook. As they had demonstrated over the last few months, at least from my vantage point, their relationship with reality seemed somewhat tenuous.

Again the solicitor told me it would cause Thomas Cook serious issues in defending the claim if my statement, as it stood, was provided to a judge. I suggested that therefore the company should have dealt with it in a way that would not have caused them embarrassment in the first place.

Unlike the remotely constructed, solicitor produced witness statement, which was woolly and contained numerous inconsistencies, my account was precise and highlighted that this episode took place years ago. I informed the solicitor I would be more than happy to answer any questions relating to the case from either the judge or opposing barrister based on my statement.

A few days later, the other captain, Garry, contacted me. He sounded incredibly nervous and stressed. He was anxious about the whole situation and how it might affect his future career. He told me he had signed the solicitor's pre-prepared, witness statement. I assumed he had done so in order not to cause friction between himself, the solicitor, Hutchings and Thomas Cook.

I asked Garry three simple questions.

Did he, at any point, see me operating outside SOP's?

His answer was 'no'.

Did we both follow SOP's to the letter?

He replied 'Yes'.

Finally, I asked him had Thomas Cook in the last three years, contacted him about any issues that day?

'No'.

Just before Garry finally hung up, he sounded more deflated than when he first rang. I had explained the reason I would not sign the original witness statement. I suspected the fact that he *had* would now leave him open to answering some difficult questions, not of his own making, other than agreeing to sign, if it was the same as the one I was sent, a *debatable* witness statement.

I had never been so popular with legal teams. Numerous further phone calls continued from the solicitors. All tried to convince me that they were there to 'protect' me. All this succeeded in doing was to give me further reason to conclude they and Thomas Cook would have no issue in throwing me under the 'air' bus. I explained my honest version was required so that there would be no opportunity for any misunderstanding by the judge as to what we did and did not do on 20 August 2010.

One great advantage of an airline pilot operating 'by the book,' as I had since 1989, was that any individual or organisation of dubious integrity or intentions, would not have a leg to stand on if they attempted to later challenge your actions.

Still, the Thomas Cook solicitors were extremely unhappy and began to pester me daily to alter the statement. After several more phone calls and emails, I finally had enough. I let them have both barrels. I emailed the following back to them:

"I assume you enjoyed your time off for Easter and are now back at work. I worked over Easter, and I am now on days off.

Let me be very, very clear. I was not aware during the phone conversation in November, as I stood in the middle of a busy airport terminal, that I was providing a legal 'witness' statement. Indeed, if I had been aware, I would have refused. If you wanted one, you should have asked me to write one, but you didn't.

I have not changed anything in my witness statement; your company produced a document and declared it my witness statement. You cherry-picked snippets from a conversation in November, missing what I consider to be key points, introducing numerous erroneous statements and asked me to sign it. This is simply NOT ACCEPTABLE!

You state 'It is difficult to evidentially deal with your assertion that you followed all SOPs. While we would obviously state that you did follow all the SOPs and we suspect that captain xxx would state the same, there is no satisfactory alternative to your providing that confirmation yourself.'

Please re-read my witness statement note 37, where I state clearly. 'What I can be sure of is that Garry, a captain, who on this day was operating in the role of first officer, and I, followed precisely directions from the air traffic control tower, the SOPs and our training exactly.'

I don't think I could have been any clearer, could I?

You state 'While we do understand that there have been changes to the SOP since the incident, it will be difficult to defend yours and the company's position around these changes'.

I totally disagree with that statement. The fact is, the company and ATC have come to the conclusion that the jet blast from one engine (which would be double that were both engines running) is perfectly safe within the airport environment. My recollection, and it is nearly four years ago, was that we had two engines running, therefore, half the thrust that has now been considered safe by the company and Manchester Airport was being used that day. I would put it to you that using that information, it would be extremely easy to defend the company. Indeed, that's why it is important and that's why I have included it. If you are saying that it is NOT safe to taxi with one engine, then we really do have a problem.

Also, I draw your attention to the following, you state 'you would be at liberty to attend and watch, (albeit without a witness statement, you could do no more at trial)'.

Let's be clear I DO have a witness statement. I have written it containing all the elements I feel are important. It's your office that does not wish to present it to court. I shall leave you to answer why you do not wish to provide this document to public scrutiny. I am more than happy to do so.

If, despite my clear witness statement, your office does not provide the court with full disclosure of all relevant documents and this results in any suspicion that either myself or Garry operated in a manner other than with the utmost professionalism and fully in accordance with SOP and ATC instructions as I have clearly indicated in my witness statement, please be aware I will have absolutely no hesitation in commencing legal proceedings."

I did not hear back from the solicitor.

A few days later, my home phone rang. It was the ex-BALPA CC chairman, now chief pilot. Scadeng informed me that he had been instructed to phone me by his superiors about a witness statement. It was concerning my comments in the statement about Thomas Cook not contacting me about the incident and the single-engine taxi. He made it clear that both he and the company required I alter this witness statement and remove those comments.

I refused.

Did I mention I had never had any disciplinary issues in my entire flying career?

# CHAPTER TWELVE

## *Now It All Makes Sense*

*"An operator shall ensure that for all its flights, flights are planned to be completed within the allowable flight duty period, taking into account the time necessary for pre-flight duties, the flight, and turnaround times."*

E.U. Ops (Aviation Law)

Roger Scadeng had moved directly from being the chairman of the union to Thomas Cooks' chief pilot without any grace period. This was not only against established BALPA policy, but also, I would suggest, any reasonable person's idea of acceptable behaviour. Despite this, BALPA HQ had refused to sanction him, or even, as far as I was aware, comment on the situation.

Many work colleagues were quite vocal that they suspected the chief pilot position was the carrot dangled in front of Scadeng by Hutchings during the alleged *'currygate'* meeting. I had no reason to disagree with my colleagues.

A number of pilots also speculated this was precisely why Scadeng, as BALPA CC chairman, had moved from rejecting the companies proposals for severe cuts in salary, redundancies, demotions and forced base changes to recommending the pilots vote in support of Hutchings demands. They also concluded that it influenced the recommendation that the pilots accept only a fraction of the shares they were promised in the pay deal. Again I had no reason to come to any different conclusion to that expressed by many colleagues.

Virtually every day off, pilots were contacted by the Thomas Cook crewing department and pleaded with to come to work as the company now had an '*unforgivable lack of crew*'. This unforgivable lack of crew was down to one man. Hutchings. In Bristol, aircraft were grounded. Thomas Cook's excuse? Pilot sickness. The real reason? The pilots had run out of duty hours. There were several pilots prepared to capitalise on the misfortune of colleagues who had been demoted, made redundant, or kicked out of base. It was always the same faces who would agree to work days off and earn a considerable amount of extra blood money. Looking after number one seemed to be their only concern. As for myself, despite the daily lure of around £1000 to work a day off, from the moment Thomas Cook caused chaos to so many lives, I declined to sell my days off. It did not sit right for me to profit from others' misery.

Despite the pilots who were mopping up the day off payments, the airline was still critically short. It had got to the stage that Thomas Cook was having to sub out flights at the cost of millions. Crews were being flown all over the UK in private jets so that they would be within hours to do whatever duty was planned. The wheels were coming off this particular wagon. You did not need to be Steven Hawkins to work out that this situation was directly attributable to Hutchings savage cost-cutting plans. It wasn't just money Thomas Cook were losing, it was also their reputation.

Like every other single pilot, I was getting hammered. I have no idea how those who were selling their days off could claim they were not suffering the effects of fatigue!

It was clear, we simply did not have enough crew to cover the planned operations, and this summer was going to be brutal.

After six months operating out of Birmingham, I was now back at the Manchester base and had a much-needed couple of days off before my next run of duties. The first of which was a standby from 06:00z until 12:00z. That meant at any time within that period the company could call me to operate a duty. There was a time when a standby was used for the purpose it was designed for. It was an extra couple of pilots available to cover such instances as another pilot becoming ill, crew running out of hours, technical problems etc. Standby now meant being used to firefight the '*unforgivable lack of crew*'. The reality was, at some point, I would be needed on 3rd May with it being an odds-on favourite that my phone would ring at precisely 06:00z to drive the 1 hour 15 minutes commute to work. The airline demanded that all crew live within ninety minutes of the airport, to ensure any call out at short notice would not result in a significant delay to a flight.

With the potential that my alarm call on the 2nd May would be a phone call from crewing, I decided to get to bed early. Before retiring, I checked my future duties via my roster, which was hosted on the company's web portal. If the company knew in advance I would be needed, they would adjust my roster accordingly.

At around 20:00z, I flipped open my laptop to be greeted with the unfortunate news that an African general had died. His relatives informed me that as I was such a principled individual, they wished to deposit £10 million in my bank account to look after it for them until they escaped to a friendly country. Just thinking of the interest that would generate, gave me a warm feeling. The other bit of hot news my laptop delivered was that I was needed to fly the next day. Crewing had given me a report time of 12:50z, to then position, by road, to Glasgow, a four-hour car journey. Upon our arrival at the airport, we would then have an hour to pre-flight the aircraft and then position an empty Boeing 757 back to Manchester.

Well, at least I would not have to endure the rude awakening of my mobile phone tomorrow morning and suffer the indignity of going to bed before my teenage children. Such is the life within the highly glamorous world of the jet-setting airline pilot.

The following morning, I made the most of the luxury of my internal body clock deciding when I would wake. After a leisurely breakfast, I set off for the airport, two hours before my report time, which would allow for any traffic problems that morning on my side of the road or to factor in the rubber-necking goons gawping at any accidents on the other side.

Although I would only be flying for around fifty minutes, I would not be getting back home much before 21:00 that evening. Not that I was in a rush to get back tonight. That evening there would be several ladies descending on my house as my ex-wife, was hosting a 'hand-bag' party with Lisa and some friends. I was missing a treat, but this was a burden I would bravely endure.

With about twenty minutes to spare, I arrived at the crew room and began printing off the documents that were needed, even for such a short flight. Flight plans, weather, loading information, they all came churning out of the printer. I was on a roll. It usually took a few muttered obscenities to coerce the connected hardware and software to produce anything useful.

A few minutes later, the first officer arrived. At this point, we commenced the age-old ritual that has taken place for every single sector duty period since the Wright brothers first flew. Who was going to fly the aeroplane and who was going to be the secretary dealing with the paperwork and answering the phone, sorry, liaising with air traffic? It is a well-rehearsed, well-choreographed, vocal exchange, similar to the one involving the last chocolate biscuit left on the plate.

The captain starts the dance,

'Do you want to fly it today?'

The correct reply from the first officer should be

'I'm not bothered; do you want to fly it?'

Other acceptable responses rarely offered from a first officer are,

'No, I prefer doing the paperwork and having you fly it.'

Or

'I would very much appreciate the opportunity to observe your legendary aviation abilities, to hopefully be able to advance my knowledge in the hope that one day I will be able to emulate your prowess.'

The absolutely totally incorrect answer is,

'Yes!'

That answer would immediately result in adding that particular first officer to the admittedly short list of those of whom the question would never be asked again. Assuming one of the correct answers was offered, we then proceed to the next part of the dance. We now each fumble in our own pockets to produce a coin which will make the decision for us. I understand that the British Airways legacy contract pilots have a similar procedure but with an American Express platinum card. It is vitally important at this stage to ensure that both parties understand that this is a single toss of the coin or the best of three or five.

As usual, I lost and, as he was an experienced first officer, he would not allow an extension to the best of three.

I had plenty to do on the 4-hour journey up to Glasgow. Every six months, airline pilots must undergo rigorous simulator sessions held over two days, and it was my turn for the 'box'. As I would be sitting in the back of the taxi, I decided to make the most of the time by reviewing the systems of the Boeing 767. The simulator sessions are designed to enhance the knowledge of the aircraft in a safe environment. It is also to test and ensure a pilot can meet the required standard of knowledge and still exhibit the required flying ability. Scenarios that would be dangerous in a real-world environment can be examined. How you deal with failures such as an engine catching fire, or hydraulic or electrical issues can all be assessed. Other scenarios that are dealt with include bad weather that necessitates a 'go around' at the very last moment, all the while flying with only one engine. Other possible scenarios are failure of navigational aids, fuel leaks, wind shear, rapid decompression, fire on board, or any combination of the above. Many other problems can be dreamt up by the training captain sitting in the back seat of the simulator, usually accompanied by an evil grin.

Although a significant aspect of the two days is training, there is a testing element, and the Thomas Cook training captain, Tim Cheal, had made it very clear. All simulator rides were 'jeopardy' rides. That means a failure of any single section will require that section of the test has to be re-taken. Any inability to reach the required standard, for whatever reason, can ultimately result in being demoted or your contract being terminated with immediate effect. I had witnessed the effect this added stress can have on a normally proficient first officer several times. Unfortunately, within the world of aviation, you're only as good as your last simulator check. Fortunately, during my time flying the Airbus, my simulator rides had always been to a high standard. Now, however, I was flying the Boeing 757 and 767, and this was the first time I had ever flown a commercial airliner without the benefit of a year or so flying as a first officer and gaining valuable experience. I had, therefore, missed out on the opportunity to observe how an experienced commander would operate the aircraft. I had to learn as I went along. As most professional pilots would admit, it is personal pride in your performance that drives the preparation for a six-monthly check. The next four hours in the taxi would therefore be time well spent.

Before I knew it, we were pulling up outside Glasgow airport terminal. Fortunately, today the terminal was free from religious nutters, and no one was trying to ram a four by four into the terminal. All we had to face now before reaching the comparative calm of the Boeing cockpit, was Glasgow airports own security, guaranteed to sort the men out from the boys in the blood pressure department.

Finally, after having left the house around eight hours earlier, while the first officer commenced the walk around, I started the internal checks, ensuring everything was in the right place before loading up the aircraft's navigational computer for the very short flight back to Manchester.

All memory items completed, we commenced the 'before engine start' checklist, which includes phones being switched off or put into flight mode. Literally, as I searched for the flight safe mode page, my phone rang. It was Lisa to tell me that my ex-wife, my children's mother, had just been rushed into hospital and things were looking very serious.

The flight itself was uneventful, but as soon as the shutdown checks were completed in Manchester, I rang Lisa for the latest update. My ex was now getting the best care she could while the doctors attempted to stabilise her. I eventually walked through my front door at around 21:10z. It was a 10hr 20min working day, but my attention was firmly focused on the medical situation and its repercussions. The primary attention was to ensure we could assist my ex in any way that we could and to ensure the children did not worry too much. Fortunately, her condition stabilised the following evening but, she remained in hospital.

The next rostered duty was a report at 04:15z in the morning for a flight to Palma Majorca. It would have meant waking up only 4hrs 35 minutes after I had walked through the door. Fortunately, legislation is in place to prevent that from happening. The absolute minimum rest time between duties is twelve hours. If the previous duty were greater than twelve hours, then the length of that duty would become the minimum rest period. The late finish meant that I would not be waking up with the undead and flying to Palma and back. My day would instead revolve around doing my best to mitigate the impact of the sudden medical situation with the children while preparing for the next two days in the simulator.

Both simulator sorties were planned for silly o'clock. It was appalling planning by Thomas Cook crewing. Shifting from an evening duty back to very early duties was never a smart idea, but with the "unforgivable lack" of crew, what else could they do?

I decided to go to bed the evening of the 3rd of May at 19:00. Three alarm clocks were primed to wake me at 02:00z. I could not even risk setting them 10 minutes earlier in the hope of enjoying a ten-minute snooze. This would leave far too much scope for Mr Murphy to enact his law. Fumbling to reset buttons and switches in a near-catatonic state was just asking for trouble. That's the problem when you wake up suddenly.

So, here I was, in bed at a time only known to naughty school children, trying to fall asleep two hours before the time I had walked through the door the previous evening. The more you try to sleep, the more your brain reminds you that you're trying to get to sleep. The more you try to rid your mind of conscious thought, the more space you leave for the realisation of exactly how much your mortgage is going to cost you after 25 years paying the bank.

This particular problem is technically known as 'evening week maintenance zone', or, as it's more commonly known as 'I can't sleep because it's too early to go to bed.'

Compound interest and APR seemed to do the trick, and at some point in the proceedings, I drifted off until the evil arpeggio of three alarm clocks sprang into life at precisely the same time in a bid to warmly welcome me to the wonders of what the day might hold.

Up at 02:00z and after the obligatory spectacles, testicles, wallet and watch check, I was in the car for my hour and a half drive to the simulators based in Woodford. Of course, neither the company nor CAA make allowance for the extra time it takes to drive to Woodford for the simple reason, they do not legally *have* to, and if you don't legally have to, why would you?

For those of you not familiar with an aircraft simulator, it resembles a large box on legs, with a tap on the side that allows the sweat generated by the two pilots under examination to be safely removed. It is, in effect, the most realistic video game ever released, only in this case, 'game over' means you will be looking for another profession to feed your wife and children.

A simulator session consists of a one-hour pre-simulator brief. This is then followed by two separate, two-hour sessions where various failures and scenarios are thrown at the crew. The two sessions are separated by a fifteen-minute break to allow you to replenish your fluids, or alternatively, to get rid of some.

Fortunately for both myself and the first officer, the gods were smiling on us, and we both passed the first day’s run of failures and approaches. Although we were only at the training centre for six or seven hours, those hours were crammed with more problems than most pilots experience in a lifetime. The effects of stress and its impact on pilot fatigue is well-documented, being much more significant than experienced during a regular flying duty. Unless it is really not your day, it is highly unlikely you would have to deal with engine fires, a fuel leak or other such thought-provoking failures on your way to Malaga. In a simulator, those problems are only a single press away on a touch screen in the hi-tech box.

Having left at 02:15z, I was finally back home and sitting in my kitchen by 12:15z. The rest of the day was spent continuing to deal with repercussions of the medical situation and preparing for the next

Groundhog Day in the simulator, where only the scenarios would change. I was again in bed at approximately 19:00z, this time falling asleep after having mulled over such important cerebral issues as, is Kanye West's real name Ken, and what exactly do the Kardashians do?

02:05z the next morning, still half asleep, I'm once again drinking copious amounts of orange juice to kick-start my system. I had never been a fan of coffee.

The music of choice for the drive this morning would be Rammstein. Most music lovers will be aware of the Teutonic group. Still, for those of you reading this book who consider The X Factor having anything to do with music, Rammstein is a German oompah band known for their stylish choice in lederhosen, felt hats and pan pipes.

By 12:15z, I was once again walking back through the front door, but this time with my pilot's license containing the training captain's signature. This would allow me to continue to fly the airline's passengers for the next six months. At least that was the plan.

Through the very divisive recent change in the pilot's terms and conditions, the company could now change the original start time of any duty that had been rostered by up to four hours. They could also change the initial rostered finish time by up to four hours. The upshot was, the airline could shift the working day by up to eight hours.

The effect this had on every pilot's roster was that it became as much a work of fiction as an MP's expenses claim form. In reality, the only time it was worth checking what you would be doing the following day was immediately after you finished your current day. This was the case when I returned home after the simulator duty on the 5th of May.

The duty of May 6th had initially been a standby. Due to the *'unforgivable lack of crew'*, the standby had been changed weeks before to a flying duty. For reasons already explained, I had not taken much notice of the change apart from the fact that I knew I would now be operating.

The next day's duty would require yet another early start, with the two iPhone's and iPad chorus again waking me with a comparative lie-in until 04:30z. The report for the duty at the crew room in Manchester was 07:00z. As usual, the required one hour for pre-flight duties was included on the roster.

The roster showed that we would depart Manchester at 08:00z to arrive at Heraklion, Crete at 12:00z, a sector time of 4 hours. The roster then showed a turnaround time of one hour. The standard turnaround time however for the Boeing 767, which was the aircraft that I was rostered to fly that day is 1hr 15mins, as detailed in the company's own manual. After that one hour turnaround, we were then to leave Heraklion at 13:00z with the roster showing an arrival time at the next destination which was Antalya in Turkey of 14:05z. Yet again, 15 minutes were shaved off the standard turnaround time, with the roster showing a departure time from Antalya 15:05z. Very importantly, the arrival time displayed on the roster back into Manchester was 19:30z after a sector time of 4hrs and 20 mins.

For those unfamiliar with aviation terms, a sector time starts from the moment the brakes are released, and the aircraft is pushed back off stand. A sector is completed when the aircraft arrives on stand and is chocked, at the destination. The sector time accounts for the period taken to taxi out to the runway, the take-off, the climb, the cruise, the descent, the approach, the landing, and of course the taxi back onto stand.

Having looked and absorbed the roster that the company had produced, I was concerned. There was that stench again.

Most will be familiar with the principle behind the tachograph. It is a requirement for operators of all heavy goods vehicles to ensure that their vehicles on the roads of the UK are equipped with one. It is a device that monitors when a driver is driving and when he's resting. The reason for its existence is simple. It is to prevent unscrupulous haulage companies pressuring its drivers into driving beyond a period that has been deemed safe. It protects and prevents a family with two children from being ploughed into by a twenty-tonne truck carrying a passenger in the form of a driver who is asleep at the wheel.

The tachograph information is binary. There's a specific point in time that the driver is legal, the moment after, the driver is not. The driver and haulage contractor are both liable, and the penalties are severe if the maximum time is breached. Of course, there are exceptions to this situation, but this is only allowable for reasons to ensure the safety of other road users. You would have therefore thought, with the responsibility that comes with flying hundreds of people in a machine that weighs 230 tonnes travelling at .82 Mach, (that's .82 the speed of sound, or around 500mph in old money), there would be a version of the tachograph?

There isn't.

There is, however, a limit to the maximum flight duty hours any pilot might fly on any particular day. It is imaginatively referred to as the maximum 'flight duty period' or, as we are operating in the aviation world, and pilots love abbreviations, the max 'FDP'. The max FDP is based on several factors. The first is the local time of the report. It then takes account of how many sectors will be flown that day. Finally, how many pilots are in the cockpit. This then becomes the absolute maximum time that an airline can plan for a series of flights. This is made very clear within the legislation that covers every airline that is based within Europe.

The specific legislation states:

"An operator shall ensure that for all its flights, flights are planned to be completed within the allowable flight duty period, taking into account the time necessary for pre-flight duties, the flight and turnaround times."

Failing to comply with the above is punishable with up to two years in jail and a significant fine. It is a breach of statute. It is a criminal act!

The critical words within the legislation are "ensure" (definition of ensure is - to make sure, to make certain, to *guarantee*.) The next important word is "all", meaning, er... 'all'. The vital phrase is "planned to be completed".

Again, as ever, there is an exception to the legal maximum FDP. As anybody who has operated within the world of aviation for any period knows, things very rarely run smoothly. Even if an airline had complied fully with the legislation to plan a flight within the maximum legal FDP, sometimes actual events, *after a report time,* can conspire against them. By the time the aircraft lands on the final sector, this might well result in the pilots and crew having exceeded the original maximum FDP. This may occur due to several reasons, typically significant weather, technical issue or diversion that happened or became known about *after report.*

So, what if, *after report,* something like this happens and causes us to exceed the maximum FDP? Does it mean that because the fuel truck is late, or five passengers get lost in the terminal or any other unforeseen problem that delays the aircraft thirty minutes, that three hundred passengers and crew are now stuck down route until the next day? Actually no, it doesn't.

For unforeseen delays or problems that occur *after the duty has commenced,* the commander of the flight can, if he so chooses, extend the maximum FDP. This extension may be up to three hours. This is known as commander's or captain's discretion. Once again, the hint is in the name. It is solely at the discretion of the captain. He can decide to either exercise that discretion or not. Airline management has no say whatsoever. This is extremely important. It supposedly prevents unscrupulous managers from taking action against a particular captain for refusing to extend the duty.

Captains are tasked with making the right decision. That's why they are paid the salary they are. Any decision, for obvious reasons, should be free from any real or perceived pressure from any airline that wishes to discount flight safety in favour of commercial expediency. The use of captain's discretion is also carefully worded in the applicable documentation. It states:

"An aircraft commander may, at his discretion, and after taking note of the circumstances of other members of the crew, extend an FDP beyond that permitted in paragraph 13, Tables A, B, C, or paragraph 23, Table D, provided he is satisfied that the flight can be made safely. The extension shall be calculated according to what actually happens, not on what was planned to happen."

It is therefore clear that command discretion, even if the captain wanted to, cannot be used to cover any failure of the airline to ensure, at the planning stage, that the series of flights can be completed within the maximum legal number of hours.

With the above in mind, and having already been awake since two this morning, I now had to consider what tomorrow's duty would entail.

With a report time of 06:59z, the maximum legal duty period would have been 11 hours 30 minutes. However, with a report time of 07:00z, legislation allowed a maximum legal FDP of 12 hrs 30 minutes. For want of one minute, the airline was able to require its crew to work another hour, but, the line has to be drawn somewhere, and the line was drawn at 07:00z. Reporting at 07:00z with a legal maximum FDP of 12 hrs 30 minutes, via simple deduction dear Watson, meant that in order to prevent those that planned the duty having to be careful when picking up the soap for two years, the aircraft must be back on stand at Manchester by 19:30z. The airline, according to the law, had to *ensure* it.

Looking at the roster, well, surprise, surprise, it showed being back on stand at, yes, you are ahead of me, 19:30z!

The roster is compiled by taking the historical times between city pairings, and, using the data collected over many years' adding them together to give the crew an idea of what time they would be back at base. The hint that the roster is not the 'plan' is based upon the fact that it's called a 'roster'. Likewise, the indication that the documents issued on report is the 'plan' is due to the word 'flight plan' emblazoned on it.

The CAA themselves define what a roster is.

"A duty period, or series of duty periods, with stipulated start and finish times, notified by the operator to the crew in advance."

I can see reference to a *duty period,* but no, as hard as I look, I cannot see any reference to a *flight duty period.* These are two very separate things.

My first question therefore was, had the company ensured, guaranteed, made sure, as is required by legislation, that the FDP would be completed within the legal maximum FDP? My second question was if they had not ensured this, if when the flight plan was issued, it showed the aircraft landing back into Manchester after 19:30z, could I legally use captain's discretion to cover for that failure?

My answer to both questions was no.

My next predicament was, what do I do about it? Do I do nothing and turn up for work tomorrow at 07:00z, mutter the Harry Potter incantation and wait for the flight plans to be printed. Then and only then, when I had the flight plans, the evidence, in my hand, inform Thomas Cook that the FDP was in breach of legislation?

If I was a betting man, based on sixteen years of flying for the company, it was an odds-on cert that the flight plans would confirm the FDP exceed the maximum time and would therefore be illegal. If I did leave it until the next morning, that would drop the company right in it. I could not legally use my discretion before we left to cover the illegality. The breach would have been known about before report time. Therefore, they would have to come up with some other plan, at the same time as having an *'unforgivable lack of crew'* .

I was, however, now 'off duty' and there was no requirement for me to give them any head's up of the virtual certainty that tomorrow's duty would cause chaos. In all my time with the company, I had never 'dropped them in it', and I wasn't about to start now. At the very least, it would be the professional thing to do to give them the advanced warning of the certainty of the issue.

In order to prevent the inevitable, I phoned Thomas Cook's crewing department and spoke to a guy called Joe Cavanagh. I explained that I did not believe the schedule's allotted time was going to be sufficient to cover the legal requirements of the FDP. I offered several solutions that would prevent this from happening. I suggested putting another crew on the aircraft so that myself and the first officer would fly the first leg to Heraklion and then onwards to Antalya in Turkey. At that point, the new crew would fly the aircraft back to the UK. The result of this common fix to the issue would be that both crews would have operated legally. Another solution was that an extra pilot could be put on the flight deck. This would have increased the maximum allowable FDP by an hour. I actually then realised that particular idea would not fly. This fix was only applicable to long haul sectors. Joe rejected every single one of my solutions to the problem. Exasperated and exhausted, I then suggested they swap my duties with another captain and put me on a later flight that maybe was also a shorter duty day.

I told Joe that the company had also shaved 15 minutes off the standard turnaround time and I had no doubt that the flight times of the three sectors had been falsified to make it appear that the duty would fit. I was clear; I would not be utilising my discretion to cover any falsified duty times. It would be illegal to do so. Thomas Cook's own documentation stated it was the airline employee's responsibility to highlight any fatigue risk. Considering the switching back and forward between late, then early and then back to very late start times and the length of duties that week, I highlighted that the series of flights would very likely, ultimately, result in fatigue. I had offered to be flexible to ensure the flights were operating legally, but far more importantly, safely.

Every one of my suggestions was refused.

After the call, I considered just the last leg of tomorrow's duty between Antalya and Manchester. Experience indicated that for virtually every flight, according to my logbook, the sector time was in the region of 4 hours 45 minutes. Yet, according to the roster for tomorrow's flight, the company had only shown 4 hours 25 minutes. Basic maths meant you did not need to be Inspector Poirot to work out what they were up to and that it was a certainty that when tomorrow's flight plans were issued, they would be in breach of the legislation by at least 20 minutes. They would then expect me to cover that breach by illegal use of my discretion.

Unless I was very much mistaken, it looked like I was going to have to draw my own line in the sand tomorrow. Ultimately, however, that's precisely what I am paid to do, and exactly what an airline that places safety and legality before any other consideration would *expect* me to do.

Unless, of course, the CAA has concluded it is legal for an airline to falsify and fabricate flight times to *'make it fit'*. I guess I would find out soon enough.

# CHAPTER THIRTEEN

## *Up To The Barbers Pole*

*"A crew member shall not operate an aeroplane if he/she knows that he/she is suffering from or is likely to suffer from fatigue or feels unfit, to the extent that the flight may be endangered."*

E.U. Ops (Aviation Law)

For the third time in three nights, I was in bed before my children, and whatever happened tomorrow, I knew it was going to be a long day. In preparation for the next attempt at sleep, I began the ritual of setting the alarm clocks, but this time not for 02:00z like the previous two nights but 04:30z. I could almost set them in my sleep by now. At this rate, I might have to.

I was rudely awakened at the appointed hour, and by 06:30z I was driving around the staff car park at Manchester looking for a parking spot.

I eventually walked into the crew room ten minutes before I needed to be there. The first officer today was Phil. He had already waved his wand and uttered the required incantation. The flight plans already in front of him testified the printers and computer terminals were functioning. The flight plans proved that my E.S.P. abilities were as sharp as ever. My prediction the previous day that the FDP would be in breach of the legal maximum number of hours was correct. The flight plans showed landing back into Manchester twenty-three minutes beyond the maximum allowed. The flight plans had been calculated and produced early that morning by Thomas Cook at 06:01z. It was clear that, as it stood, we could not operate the duty. The law dictated we must be 'on-chocks' by 19:30z. The plans showed it would be 19:53z. There it was, in black and white, on a legal document that the airline had produced but required my legal signature of acceptance.

I described yesterday's conversation to Phil. I told him that not only had I warned Joe about this, but also I would not legally be in a position to offer command discretion. Phil, a very experienced first officer, agreed that, as it stood, we could not leave Manchester. We were stuffed!

Was this situation due to a barrow boy culture exemplified by Hutchings' apparent inability to stick to agreements now oozing its slimy way down to certain individuals within the crewing and rostering departments? As far as Phil and myself were concerned, the flight plan printout was evidence of their failure to abide by the law. In producing a flight plan before report that exceeded the max FDP, Thomas Cook and those involved had, in effect, provided *"the evidence for their own prosecution."*

This put Phil and myself in a completely avoidable bind. I suspected this was precisely what Thomas Cook was banking on all along. The proof that they were expecting us to just 'get on with it' was staring back at me. The plan was seducing me just to accept this little illegality, this ever so small morsel of criminal activity. *'Go on, do it, no one will know, no one will say anything. Nothing terrible will happen; everyone else would do it. Come on, it's only twenty-three minutes.'*

Would twenty-three extra minutes make any difference to the safe operation of the flight? Highly unlikely. But remember that for want of one minute, the absolute maximum duty would have been one hour less. We would now be talking eighty-three minutes in breach.

Try falsifying the tachograph on an HGV by twenty-three minutes and see how long it takes before you had your collar felt by Mr Plod.

If you accepted a twenty-three minute breach, why not accept eighty-three, two hundred and three, hell why not do away with the limit altogether and claim as long as your eyelids were open, it was legal?

The point was, lawyers had drawn a line in the sand so that a pilot would not have to. It appeared, however, Thomas Cook were now nonchalantly walking along, hands behind their back, whistling a happy tune while looking up into the sky. All the while their feet were now busy attempting to erase that sandy line. It looked like I was going to have to get a big stick and redraw it for them.

Our one hour's report time was for checking weathers and for any potential issues en-route at the destination or diversion. This information would inform your final decision to the required fuel load. What the hour report was not designed for was to deal with this bullshit.

Courtesy of not being listened to by Joe, we now had the extra workload of having to find a solution to a problem not of our making. Alternatively, I was entirely within my rights to say 'told you so', turn around, and hop back on the staff bus.

The pressure to comply with this illegality was enormous. Conflict with the management and several hundred passengers having their holiday plans severely affected due to a situation that should have and could have been resolved the day before loomed.

It would be times like this that I would recall my father's comment about how the industry would likely conspire to ensure that it was the pilots who would be blamed in the event of an accident or incident. What would be the first thing any investigation would try and uncover? Find any weakness in the pilot's operation or failure to abide by all rules and procedures.

On that basis, let's say because I was a senior captain at the company, and I did not want to be seen as *'difficult'*, I accepted the illegal flight plan. Imagine if during the day something had happened. I would be hung out to dry. I would have knowingly operated illegally. If I accepted an illegal duty, I would have put my neck in the litigation noose.

Having discussed all the options, which by now were quite limited, we rapidly concluded that the only way this duty could be flown legally would be to divert on the last leg between Antalya and Manchester into Gatwick. The aircraft would then be on chocks at Gatwick before the witching hour. Considering it was still before 07:00z in the morning, this would give the company over ten hours to source two pilots from one of the airlines main bases, or at a pinch, position a crew from another station. This replacement crew would then fly the aircraft on its last leg to Manchester. If they could not find two pilots in twelve hours, you would have to seriously question Thomas Cook's ability to run an airline.

This would not be anything out of the ordinary. Thomas Cook would regularly advertise in its brochures a 'direct flight' to a holiday destination. They would do this knowing that, given the distance between the two points and the expected number of passengers and bags, the aircraft simply did not have the range. On the day in question, the plane would depart in the full knowledge that an 'en-route' diversion would be required. A suitable airfield would be chosen and would be used to refuel the aircraft. This extra time would be known about days, if not weeks before the day commenced yet it would not appear on the roster.

I rang the Thomas Cook crewing department and spoke to a Martin Lamb. I explained the situation to him, and, as I could not use my discretion, asked 'where would you like us to divert to?'

Lamb confirmed that Gatwick would give the company the best chance to find a replacement crew. That seemed a sensible idea.

I then rang the airline's operations department and spoke to Peter Lloyd. The operations department's responsibility is to deal with all the issues that ensure an aeroplane and its pilot are in the right place at the right time. Considering the aircraft, its pilots and crew and over three hundred passengers would still be at FL 360 at 19:30z, something had gone seriously awry. Again I highlighted I had predicted this situation the previous day. I then began to explain the Gatwick plan, and the reason I could not use my discretion as the plan had been planned illegally in breach of the maximum hours,

Lloyd interrupted me:

'It doesn't really matter what you say to me. You need to tell crewing.'

Well, so much for the teamwork the company kept banging on about.

This brought us to the heart of the situation. Captain's discretion can only be used for *'unforeseen'* events, *after report*, such as a problem with the aircraft or for weather issues that might cause that day's duty to exceed the planned FDP. It was not there to cover impossible flight times on a roster. Let's be under no illusion. It appeared Thomas Cook had attempted to give the misleading appearance, via the roster, that the FDP on the day would be completed legally. If they had done this in the belief I would not pick up this breach or, in the interests of not creating career jeopardy, would agree to use discretion illegally, well, they were barking up the wrong tree. Whatever the reason, they had been rumbled.

This morning's situation, by definition, was foreseen. I had foreseen it the day before.

Finally, it was agreed that London Gatwick was indeed the most favourable airport to divert to on the return leg from Antalya, and that Thomas Cook would source the required crew. It gave the company at least ten hours to source the two pilots.

We had to trust the company would arrange for the required crew. Any failure would result in the airline staring down both barrels of a financial gun. Due to the enormous amount of compensation that is required to be paid to all passengers arriving later than three hours at their final destination, it was inconceivable the company would not honour this plan. If we did have to divert to Gatwick, and there were no pilots to continue onwards to Manchester, not only would the passenger compensation be in the hundreds of thousands of pounds, but it would also mean a Boeing 767 would be at the wrong airport. Any flights planned the next day for that aircraft out of Manchester would have to be subbed to another airline. Once you then factor hotel rooms or coaches for the passengers to continue their journey that night, the consequences of not finding two pilots would cost Thomas Cook dearly. The likelihood was so remote as to not even be worthy of consideration. This calamity was why I had highlighted the problem the previous day. As they say, it was a 'no brainer' decision.

Without the agreement to provide the extra crew, it was simple. I would not be leaving Manchester Airport and, until they found another captain, neither would a Boeing 767 with cabin crew and passengers.

Another suggestion we made was for Thomas Cook to produce a new set of flight plans but this time planned at the maximum speed that the 767 could physically fly at. This would come with a massive attendant increase in fuel burn and therefore cost. The intention was to see if, what I thought at the time was twenty-three minutes in breach, could be shaved off the resulting FDP. It was just another idea, a futile attempt to get the company out of a hole of its own making. We were by now just grasping at straws. Realistically it would only save a few minutes.

In my considered view, what we were confronted with that morning was a direct result of a culture that had spread like cancer throughout Thomas Cook. There appeared to be a creeping acceptance of rule-breaking and risk-taking based on a principle that 'we got away with it once, twice, three times.' Ultimately the outcome is always the same. After the all-too-predictable, eventual, tragic event, a wide-eyed, startled, middle-lower manager is dispatched in front of TV cameras to utter the meaningless phrase, at least to those who are dead and their families, 'lessons have been learnt.' Translated, that means, 'we got away with it for years until our luck ran out.' This is the very definition of the *'normalisation of deviance.'*

Very tellingly, during both telephone conversations that morning, there were no arguments with either crewing or operations. There was also no attempt by either of the two individuals to claim the duty was legal.

By 7:15z a new batch of flight plans was issued. They still showed that we would be in breach of the maximum flight duty period by thirteen minutes. The cost of shaving those ten minutes off was immense both financially and environmentally. The amount of extra fuel burnt would be over two tonnes, and that was just for the first sector between Manchester and Heraklion.

In an attempt to teach grandma to suck eggs, the operations officer suggested that we request direct routings, and attempt to reduce the turnaround times below the hour. Bear in mind the company had already shaved 15 minutes off the standard turnaround time for a Boeing 767. We agreed to do our absolute best, but we confirmed if on the last leg the on-board computer showed we would be landing later than 19:30 we would have to divert.

In the rush that morning, I had missed something fundamental. It confirmed this falsification of the FDP was intentional. The standard time for a Boeing 767 to complete a turnaround when landed was ninety minutes for long haul operations. For short haul, the plan required seventy-five minutes. On the roster, this had been reduced to 60 minutes. However, what the company had done on the flight plan was to reduce these turnaround times even further. To cut a long story short, the illegal breach was not twenty-three minutes, but fifty-seven minutes.

Due to my continuing lousy luck in the coin toss department, Phil was going to fly the first leg to Heraklion in Crete. I would then fly the second leg from Heraklion to Antalya in Turkey and then Phil would fly the last leg into London Gatwick. At that point, we would vacate the hot seats at the front and retire to the cabin for the final short hop up to Manchester.

After ordering the fuel load, we left the crew room, and within ten minutes of passing through security, we were sitting, enveloped in the brownest of all cockpits, the Boeing 767. There was a reason for this particular interior design choice by Boeing. The brown colour scheme was chosen based on surveys with pilots claiming it was not hard on the eyes and that it helped conceal dirt.

The cabin crew had beaten us down to the aircraft. I was aware that several of the team had young families, so I explained the situation. As experienced crew, they were grateful for being kept in the loop.

As it was Phil's leg, it was his responsibility to enter into the aircraft's navigational computer the route, the expected flight levels and the expected wind direction and strength. The cruise speed was also entered, which, according to this high-speed plan was the absolute maximum speed of a 767. The maximum speed of the aircraft, for any particular set of circumstances such as altitude and outside air temperature, is represented via the Air Speed Indicator (ASI) by a red-and-white needle known as the barber's pole. In the cruise, this would sit at around Mach .82. Mach 1 represents the speed of sound. Although the aircraft itself would be flying at Mach .82, the airflow over the top of the wings would be approaching the speed of sound, leaving not a particularly comfortable buffer.

The original roster that I had challenged the day before had shown a flight time from chocks off to chocks on from Manchester to Heraklion as precisely four hours. That was supposedly using standard speed which was Mach .78. Yet here we were at the absolute maximum speed of the aircraft, and it was still showing it would take four hours and five minutes. We even had a tailwind of twenty-four knots (about thirty mph.) It was clear that the flight duty times on the roster belonged in the fiction section of the library.

With the passengers boarded, the fuel on, and start clearance received, the Boeing 767 began its journey to the other side of Europe by going backwards for about two hundred meters as we started both engines. We started the pushback precisely on schedule at 08:00z.

We had missed most of the early morning rush, and were airborne fourteen minutes later. It was an unusually swift departure from one of the busiest airfields in the UK. Departures can be quite complicated affairs due to the necessity to keep the aircraft safely separated from the inbound traffic. To simplify matters each associated runway will have a standard instrument departure (SID). Our SID that morning was a Honiley One Yankee. It required, after take-off, to fly straight ahead for 3.2 miles, jink to the right to track 285 degrees until 5 miles from the airport, then turn left direct to the HON beacon. If we got it right as we rolled to the left, I would get a good view of the Knutsford roundabout which leads onto the M62. If we got it wrong, the noise sensors dotted around the airport would let the authorities know which would result in the airline getting a letter and a demand for a several thousand pound fine, for breaching noise restrictions.

As we settled into the climb towards HON, we were passed to the radar sector controller, where we were asked the usual question as to what altitude (FL) we could reach 50 miles before the Biggin Hill beacon. The air space around London is one of the busiest in the world, with Heathrow, Gatwick, Stansted and London city all in close proximity. It was, therefore, crucial for ATC to be aware of what flight levels the transiting traffic required and also what climb rates could be achieved. The flow of instructions from the controller was virtually non-stop to one aircraft then another. They were, however, all in English! Air traffic would read the instruction out, the crew on the aircraft concerned would read that instruction back. The controller, it seemed, did not even pause for breath. Regularly headphones would be filled by a screech. This would occur when two aircraft transmitted at the same time. It had become an art in itself to know when a conversation between a particular aircraft and the radar controller was finished. It has perplexed me to this day why a system has not been introduced that prevents two aircraft talking at the same time. The resulting squeal made any communication unintelligible. Instruction and response needed to be repeated all over again.

Finally, I saw my chance, a gap in the transmissions, the radar controller must be taking a breath, his probable purple complexion now turning to a slightly healthier red. 'Kestral 2148 request direct Konan due to being tight on-duty hours.'

At this stage, this was an unusual request and would involve extra workload from the radar controller to contact Brussels and then liaise with all the relevant sectors to enable us to get these direct routings. No one could deny that Phil and myself were 'busting a gut', including the controller's, in a joint, but vain attempt to change the laws of physics and get the aircraft back to Manchester by 19:30z.

Eventually, we reached our somewhat lower-than-usual planned cruise altitude. To save fuel, the 767 would usually cruise at 35,000ft to 37,000ft; however, today, according to the flight plan, we would be cruising at 29,000ft. This would increase the indicated airspeed at which the 'barbers pole' resided due to the much denser air at the lower level. This has the advantage of allowing the aircraft to fly at a much higher airspeed before encountering 'high-speed buffet.'

I had spent the previous 14 years on the Airbus family of aircraft, including the Airbus 330. Selecting a given airspeed, with the autopilot engaged, would result in the aircraft holding that speed, usually within a couple of knots, thanks to an excellent and modern-ish auto-thrust system. This system would gain its information via several on-board computers to sense if the aircraft was accelerating or decelerating in relation to the airflow and would make minuscule adjustments to the thrust. Today, however, we were flying a Boeing 767, a much older beast. This was a little like comparing my old Amstrad computer to the latest iMac, only the Boeing 767 was considerably older than the Amstrad. The important point was, the antiquated auto thrust system would allow the speed to vary significantly, making it quite easy for a high-speed exceedance resulting in an over-speed. This extra monitoring meant it was going to be a long day with the hands hovering over the thrust leavers ready to take off a handful of power at a moment's notice.

Having sweet-talked air traffic control along the route, we managed to negotiate several further shortcuts which saved a couple of minutes. It was only ever going to be a couple of minutes. Luck was also on our side on our approach into Heraklion. The airfield, along with quite a few other Greek destinations, does not have a large apron. Parking space becomes a premium. It was typical that before being able to land, you would be required to hold north of Heraklion and wait for an aircraft to depart that would free up a parking spot. Flying at maximum speed, getting very unusual direct routings, having no delay into Heraklion all topped off with a twenty-four-knot tailwind enabled us to land five mins ahead of the scheduled time. It was basic common sense, however, to realise if we had a twenty-four-knot tailwind while flying eastbound, we would experience a similar strength headwind that evening flying westbound.

During the turnaround, our good fortune continued. Baggage handlers turned up on time, even the fuelling trucks. Very unusually for Heraklion, our departure was not held up by arriving aircraft. The result was we managed to shave ten minutes off the turnaround time. Before departure from Manchester, there would've been absolutely no way of foreseeing these exceptionally fortuitous events.

For the flight between Heraklion and Antalya, it was apparent that the company had not planned, as promised, the next sector at the maximum speed. The expected fuel burn would, therefore, be significantly higher than the figures that we had been presented with on the flight plan. The required fuel at the standard speed was just under eight tonnes. Considering we had burnt two tonnes extra flying at the maximum speed between Manchester and Heraklion, we added two tonnes on top of what the flight plan required for the sixty-five-minute hop over to Antalya. Not only would this enable us to be on the fat side for the one hour's flight to Antalya, but would also save time on the turnaround as we would already have extra fuel.

The departure from Heraklion was always one of my favourites. As the aircraft passed the end of the runway, there was a requirement to roll sharply towards the north to avoid the high ground. It was definitely a hands-on departure.

Due to the efficient turnaround, plus the fact that we landed five minutes ahead of schedule, it meant we left Heraklion 15 minutes ahead of schedule. Also on this leg, we had a 30-knot tailwind. Once in the cruise, it was time to take stock. Even with everything going far better than could be foreseen at the beginning of the day, a quick calculation still showed that we would have to divert into Gatwick to be on the ground by 19:30. Once again, we flew the sector at Mach .82.

It was towards the end of this sector that the everyday normality of operating within the world of aviation returned. The number of aircraft landing at Antalya required us to enter the hold and await our turn to make the approach. Eventually, we arrived on stand at 14:00 which although five minutes ahead of schedule, would still not be sufficient to make up the twenty-three minute deficit, considering the last leg would be payback for the tailwinds we had so far enjoyed.

It was while on the ground at Antalya I had the first indication that today's duty, combined with the last few days' early sim slots were starting to have an effect. It was only riding the adrenaline wave of flying the aircraft at its limits that was keeping me alert.

Like everyone else, pilots do become tired at some point. It is a legal requirement that individual crew members ascertain whether or not their level of tiredness will affect flight safety. When it does, that level of tiredness is known as fatigue. ICAO definition of fatigue is:

"A psychological state of reduced mental or physical performance capability resulting from sleep loss or extended wakefulness, circadian phase, or workload (mental and or physical activity) that can impair a crew member's alertness and ability to safely operate an aircraft or perform safety related duties."

Again the law is straightforward and unambiguous. It states:

"A crew member shall not operate an aeroplane if he/she knows that he/she is suffering from or is likely to suffer from fatigue or feels unfit, to the extent that the flight may be endangered."

It is just common sense.

Although tired, I did not consider it at a level where it would affect the safety of the flight back to the UK. However, it was something that

I was aware of and would need to monitor. As the legislation states, it is the crew member who will decide whether they are fit to fly. How could anyone else possibly determine if someone was, or was not fatigued?

As had been arranged at the beginning of the day, around twenty minutes before departure, I rang Peter Lloyd in the operations department. I told him straight. We would not be landing in Manchester by 19:30z. Had a new crew been arranged? Despite it being seven hours after ops and crewing had been alerted that this was a virtual certainty, Lloyd informed me no plan had been put into place. The reason? They had been busy with other matters. More of a concern was that the very same individual I had spoken to at seven that morning now seemed taken aback that I intended to do what was agreed I would do.

Ultimately, my job as the captain of an aircraft is to ensure it is flown safely and legally. My job description was not to sort out crewing issues. Any problems arising from the lack of crew is the responsibility of the chief pilot and director of flight operations, namely Scadeng and Hutchings. These two were responsible for ensuring we had the correct number of pilots. The proper utilisation of the crews we did have was the responsibility of the crewing and operations departments.

As it was now a virtual certainty we would be diverting into Gatwick, I relayed that fact to my cabin crew colleagues. Some of the crew had lucrative long haul trips the following few days and did not want to miss out. Some had problems with childcare. Every single one of the cabin crew understandably just wanted to get back home, and it felt like I was the person who would be raining on their respective parades. I decided to make the cabin crew's life a little easier and not announce to the passengers that had now boarded of the diversion. Had I made such an announcement it was a certainty that for the entire flight the crew would have been inundated with understandably upset and concerned passengers asking about what would happen next? In reality, at this stage, not only did the crew not know, but neither did I.

Again there was no high-speed plan sent by the company for the last sector back to the UK. Once again we added an extra two tonnes of fuel to allow for the high-speed cruise. I also checked the wind component on the westbound sector. When the first two sector wind components were taken into consideration, literally, to within a couple of knots, the tailwinds on the first two sectors were cancelled out by the headwind on the final sector.

The passengers were on board, the navigational computer loaded, we had enough fuel, and were all set for pushback. We called ATC.

Instead of clearing us to push and start, we were told to hold position due to a runway change. It meant reprogramming the departure aspect of the flight plan that sat in the aircraft's computer. No sooner had we completed the update, we were informed of yet another runway change. Practice makes perfect, and eventually, Turkish air-traffic control got their act together and decided which runway was in use. We ultimately left the gate ten minutes later than the falsified schedule indicated.

Antalya is a huge airport, and, as such, it can take a considerable amount of time to taxi from the gate to the runway. Factor congestion into the proceedings and you can be talking twenty minutes or more from leaving the gate until getting airborne. This afternoon, however, we were lucky with the traffic. There were no delays.

After take-off we turned the airliner towards the north-west, again trying to reduce the flight time as much as we could by compromising climb rate for forward speed. As we were handed over from each controller, our opening introduction included 'requesting direct due to crew hours'. Again the flight plan produced showed a cruise speed less than maximum. As before, in an attempt to pull out all the stops, we were once again hurtling along at Mach.82.

It was time to take stock of the situation. We had negotiated direct routings so the flight was now as short as it could possibly be. We were flying at the maximum speed the aircraft could fly at, and the headwind we were experiencing was precisely what was predicted. We literally could not have done more.

I had told Lloyd that once established in the cruise, I would contact ops and relay what the flight computer spat out. Before I made the call, I had one last trick up my sleeve to reduce the flight time.

When landing at a busy airport such as Manchester, it is normal to fly a STAR or Standard Arrival Route. This can add several miles to an approach. It is necessary to allow the maximum number of aircraft that can be sequenced to land on a runway over a given period. Unfortunately, it also adds several minutes to the block time, all of which are required to be accounted for on the flight plan. We needed to save every single minute. To give the absolute minimum flight time when I called the operations department, I took out all of the arrival routings and programmed a direct route to the centre fix. This is a point, usually five miles or so from the runway threshold. The on-board computer took a few seconds to calculate the shortened approach before providing us with its answer. We would be landing, at the earliest 19:40z. That, flying flat out with a straight-in approach, would be the time the rubber would touch concrete. The flight duty period would still not be over. It would take, at the very least, an extra five to ten minutes to taxi from the runway to the parking gate where the aircraft would be chocked. It is the time that the aircraft is on chocks that defines the end of the FDP. The on-board computers, as you would expect, are incredibly accurate, and so it was a 99% probability that by the time the flight duty period ended, as I had predicted the day before report, the flight would be fifteen to twenty minutes into discretion.

It always had been a plan that was doomed to fail.

Having now been awake for over twelve hours, and on the back of the run of the last two previous days waking up at 02:00z, I was becoming more and more tired. Today was turning into a marathon, but it was a marathon I was still confident I could finish safely. The constant monitoring of the speed and the constant evaluating and re-evaluating of our situation was bizarrely providing the natural adrenaline to keep me alert.

Like a successful chess or snooker player, a captain is paid to think several moves ahead, and assess the consequences of each of his decisions in order not to snooker either his passengers, crew, himself or the company. The next day, after the legally allowable minimum rest, Thomas Cook expected me to fly yet another marathon duty to Turkey and back. That would be a marathon I would not be competing in. I simply would *not* be at the starting line. I would be out of hours for tomorrows duty because we were now going to have to divert.

The plan when we had walked out through the crew room door that morning was to divert if it was not possible to be back into Manchester by 19:30. The diversion would add at least an extra hour, probably two, to the journey back to Manchester. I would therefore not be able to operate tomorrow's duty due to not having the minimum legal rest.

That, of course, assumed, having had all day to do so, the company had arranged for the two pilots to take the flight onwards. If they hadn't, then we would be night-stopping in Gatwick. If that *was* the case, at the time I should have been reporting for the flights to Izmir, I would not even be at the correct airport.

I also had already considered, even by this stage, even if the computer had shown landing back into Manchester before 19:30z, I still would not be operating tomorrow's duty. When I landed, I would have a three-hour round trip between the office, home, and back. It would be minimum rest between the two duties. I estimated from leaving home to getting back today would be around seventeen or eighteen hours. Factor in the two 02:00z wake ups and two six-hour sim slots, then only a complete idiot would consider that I would be in any fit state to operate a long duty day to Turkey and back tomorrow. What else could I do? I could elect to stay in a hotel near the airport. On the face of it, that would seem sensible. Past experience however suggested otherwise. Anyone who has ever stayed in an airport hotel will immediately know why. The main issue I could foresee was that the very early wake-ups over the last three days would likely cause me to wake up early again tomorrow morning.

Taking everything into account, I considered it foolhardy to operate the next day's duty. It would be a racing certainty, even if I felt okay at the start, I would, at some point, during a flight planned to land at midnight the next day, experience tiredness that would be detrimental to flight safety. The very definition of fatigue.

At this point, as we would be landing into Gatwick. That particular bridge would not need crossing.

For long-range communication, most modern aircraft have ACARS or Aircraft Communications Addressing and Reporting System. This system is similar to texting and is clear and concise. On the Boeing 767, we had HF (High Frequency) radio. HF is everything that ACARS isn't. It's cheap, distorted and is akin to listening to a huge pan of frying bacon. Thomas Cook did not want to spend the money on ACARS. A director's bonus was far more important than clear and concise communication.

HF radio bounces the signal up to the ionosphere which refracts it down to hit the ground which then bounces it back up to the ionosphere which in turn refracts it back down to earth. This carries on ad infinitum until hopefully it finally-hits an aerial in Stockholm. Once it hits the aerial, the radio operator rings up the airline operations department. Unsurprisingly the resulting fidelity of the radio call is about as lo-fi as you could get without referring to it as white noise. This was the system I was going to have to use to inform the company of our impending diversion.

'Nice to see you, to see you nice', was Bruce Forsyth's catchphrase. 'It's good night from me, and it's good night from him', was the well-known catchphrase from the Two Ronnies. Pilots using the HF system also have a catchphrase. 'Say again.' I must admit it's not as snappy as the previous two and not as well known. It was, however, a necessity, in order to have any semblance of piecing together the distorted and fading in and out communication that was the hallmark of HF communication.

During a conversation, the signal would phase, become distorted by static. Other aircraft from anywhere else in the world would also talk over your conversation. It was therefore common for the last transmission to be repeated several times before the gist of what the other person was saying would become semi-clear.

As I was operating as secretary on this particular leg, I was responsible for all the calls to air traffic control. Although as a drummer you tend to be ambidextrous, one ability I had yet to master was talking on HF, while at the same time listening to what air traffic was saying on the other frequency. I dialled in the correct HF radio frequency and then asked Phil to take over the responsibility of dealing with air-traffic communications on the VHF Radio.

It is precisely in this situation when the primary air-traffic frequency is not monitored by both pilots when serious problems can occur. It is on that basis that any action that results in taking one of the crew away from monitoring must be kept to the absolute minimum. There have been numerous examples over the years of air traffic control losing communication with an aircraft. This is usually due to the unmonitored pilot selecting the wrong frequency when passed to another sector, having technical problems with his headset, or due to simply not paying attention. Before 9/11, you could have expected a dressing down by air traffic control once communications were re-established. Now you could expect, within a few minutes, to be flying in formation with a fighter jet which will have received unambiguous orders what to do if the uncommunicative pilot failed to follow strict instructions which could include being forced to land immediately. Ultimately, if considered a threat, you could be shot down.

Even finding the correct HF frequency is a significant endeavour. It required wading through a manual that should carry instructions to bend your legs when you picked it up. Hidden away in this manual was a chart which suggested several frequencies for different parts of the world for different times of the day. It usually took several attempts before getting a frequency that is slightly less incomprehensible than the rest.

After sending out the call 'Stockholm radio, Stockholm radio' numerous times into the ether without any response, I tried another frequency. The same call, with the same answer. Static. The third frequency finally did elicit a response from the Scandic radio operator who, as requested, connected me with the operations department via phone patch.

Apparently, I was talking to Gareth Harvey in the Thomas Cook operations department. Such was the noise and static from the HF system and the din from the two engines pushing us along at Mach .82 it could have quite easily have been Gwyneth Harvey I was speaking to.

I informed operations that according to our onboard navigational computer we would not be landing before 19:40z, therefore, to remain legal, we were going to divert. I asked was it still Gatwick that was the company's preferred option?

It was then that Harvey informed me that as Thomas Cook was 'not flush with crew' there was nobody to fly the aircraft onward from Gatwick. Harvey then saw fit to state the blindingly obvious, in that if we did divert, the aircraft would be stuck at Gatwick causing significant disruption to the operation. That was precisely why, when we had left Manchester around eight hours previously, it was inconceivable that a company the size of Thomas Cook could not source two pilots. Through the hiss and the static, I explained I was fully aware of that disruption, and that was the reason I had contacted the company the day before.

Things then just got plain silly, some might say desperate. Operations suggested that to reduce the FDP, as soon as we landed into Manchester, the company would arrange for the aircraft to be met by a tow truck, to tow us off the runway and onto stand. There was one, ever so slightly significant problem with that little idea. The flight duty period does not finish when you land. It concludes 'on chocks' at the aircraft's final parking position. The actual effect of towing the aircraft, apart from blocking taxiways, delaying arrivals at one of the busiest airports in the UK, would be to increase the flight duty period by at least another 15 minutes. This was basic stuff. All they had to do was look in the manual.

Harvey appeared desperate to cover his backside. It was as clear as day he was attempting to deflect what was behind this predicament. He tried to convince me that the reason for this situation was due to a last-minute runway change at Antalya, which according to him 'you couldn't write the script for'. I reminded him the script for this was written before we had left Manchester. In reality, the script for this farce had been written, edited, storyboarded and put into production as soon as the duty was rostered.

If you wanted to split hairs, the delay due to the runway change was ten minutes, whereas we were looking at breaching the maximum flight duty period by between fifteen to twenty minutes. Harvey's tone then changed, informing me that to be doing this 'over only 10 minutes was ridiculous,' and that I was consuming a lot of time and energy from a lot of people throughout the day. To my mind, it was clear that the company had not only falsified the flight duty period, but they had never intended to source the pilots in the first place. They had conned us.

I told him I would call back after I had discussed the matter with the first officer.

Both Phil's and my initial reaction was that we would now be spending a night in Gatwick. We had done all we could to try and make up for the illegal planning, but we were now out of ideas. Did it concern us that this was going to cost the company a considerable amount of money? Did it concern us that the passengers would be significantly inconvenienced? Of course, it did. Would we allow those concerns to affect our judgement? Absolutely not. Were we not going to do anything illegal to extricate the airline from a predicament entirely of its own making? The short answer was no. In fact, that was also the long answer.

Discretion is only available due to unforeseen circumstances and cannot be used to cover what is 'planned to happen,' only what 'actually happens.'

It is solely the commander's discretion, not the company's.

The plan presented that morning showed it had been planned to breach the maximum duty period. Therefore, that morning, before report, it was illegal to offer discretion.

But hold on. I had a thought. When we walked out of the doors of the Manchester crew room, was it or was it not the case that the plan was to divert to Gatwick and for another crew to meet us? Phil agreed that this was true. Put simply, I would have refused to even depart Manchester if this had *not* been the plan as discussed and agreed between myself and operations.

Now, if the 'plan' had been to divert and pick up the new crew but what 'actually' happened was that the company had severely let us down in not providing that crew, could we now, legally, continue to

Manchester? The law states in relation to discretion *"The extension shall be calculated according to what actually happens."* We pondered this matter for a good while. The really sad thing about this whole scenario was that both Phil and myself were not surprised that the company had pulled a trick like this. My view? They had blatantly abused our trust.

Having discussed the matter at length, we both came to the same conclusion in that, as what 'actually' happened was that the company had failed to provide the crew; we could now legally go into discretion. The next step would be to consider whether we were both fit enough to continue the flight. As previously stated, it is common sense that the only person who knows if they are fit enough to fly is the individual themselves. Their ability to make that decision has to be taken on trust. If you can't trust a pilot to make that decision about his fitness level and yet you entrust him with the lives of hundreds of passengers' lives daily, someone, somewhere, has their values and priorities confused.

Unlike my own situation, today's flight was Phil's first since a series of days off. He told me he felt fine. I informed him that I was slowly getting more and more tired but also I considered myself alert enough to continue. I also believed that the extra forty minutes or so that it would take to fly to Manchester would have very little further effect on my fitness level. It was a little like having to run another 200 meters at the end of a marathon. The impact would be minimal compared with the whole.

It was time to make a decision. I asked Phil formally 'After all we have discussed are you fit to continue to Manchester?'

'Yes' came the unequivocal reply.

Before I again faced death by white noise, I had to consider the ramifications of that decision. The first thing was to let the company know that we would continue towards Manchester. That decision, however, raised another issue. By utilising my discretion to continue to Manchester, I would now be within hours to perform tomorrow's duty, albeit only just. Given the current abandonment of common sense or consideration for their crew, I considered that there would be the expectation that I would be available to operate tomorrow's duty.

Thomas Cook's own manuals make it a requirement to inform the crewing department as soon as you suspect that you will be unfit to fly a future duty either due to fatigue or ill health. It is to allow the maximum time to re-crew any flight. The manual goes further. It states that anyone who does *not* inform the company at the earliest opportunity of their unavailability is liable to have disciplinary action taken against them.

Taken from the Thomas Cook Manual:

"It is vital that you keep crewing informed of your situation. If you are not going to be fit for the next confirmed duty, you must let them know, giving as much notice as possible and ideally by 17:00z the previous day. It is acknowledged that crew are likely to suffer sickness from time to time. It is equally important, therefore, that the crewing team is given as much notice as possible to enable them to re-crew flights in good time."

Considering Thomas Cook was not able to find a crew in eight hours, it looked like they were going to need as much notice as possible to re-crew tomorrow's flight. With that, I confirmed with Phil that he was still responsible for the air traffic communications and I again called up Stockholm Radio who patched me through to operations.

Once again it was Harvey in operations on the other end. I informed him that we were going to continue to Manchester. We had just saved the company from an enormous financial hole that it had dug all by itself.

Given that when I had tried to explain the problem to operations that morning, I was told '*it doesn't matter what you say to me, you need to tell crewing.*' I told Harvey to pass the message on to crewing that I would not be available for tomorrow's duty, but I would call crewing and explain the details when I landed.

As per my training and my command assessment, the last thing I needed to get involved in was a conversation that had nothing to do with the current flight and could be dealt with after we landed. As a trainer, I would suggest anyone who considered otherwise not to be as competent as they think they are.

Next, I let the crew know that we would be continuing to Manchester and that they would not need to worry about childcare, or losing money on trips. The rest of the flight, thanks to all of the direct routings, continued in a straight line. The aircraft eventually touching down as predicted at 19:40z. Without the aid of a tow truck, we taxied from the runway, finally arriving on stand at 19:45z, fifteen minutes into discretion. The passengers, who were blissfully unaware of the fun and games during the day, began to disembark around ten minutes later.

We finally bid a very weary farewell to the 767 about 30 minutes later.

A walk through immigration and a bus ride to the car park were next on the list followed by the old favourite of 'where the hell did I leave my bloody car over half a day ago?'

Car tracked down, I finally slumped in the driver's seat and contemplated what I expected to be, a short conversation with crewing to confirm with them the reason why I would not be flying the next day.

# CHAPTER FOURTEEN

## *Party Line*

*"To allow re-crewing of flights in good time, thus avoiding disruption to the flying programme, flight crew employees are to advise the crewing department at the earliest possible opportunity they are likely to be unfit to fly by reason of sickness or fatigue. This applies equally to standby and flying duties. Failure to do this will be investigated under the conduct procedure."*

Thomas Cook Manual

As the adrenaline and excitement generated by the day's exploits slowly drained from my body, I began to wind down. Combined with my previous run of duties, I was feeling more and more knackered by the second. However, before I set off for home, I needed to call crewing.

First, I needed to ensure that operations had actually passed them the message I would not be available tomorrow.

In twenty years of flying, I never had to do what I was about to do, but the law and the company's own manual is crystal clear. Even if you only suspect you will be fatigued for a future duty, you will be committing a criminal offence if you agree to operate. I reached into my flight bag to pull out my trusty alarm clock that doubled as a phone. As usual, I had to spend the next 15 seconds of my life listening to a recorded message *'Thank you for contacting the Thomas Cook operations and crewing information system. Please be aware all calls will be recorded for training and quality purposes.'*

My concern was, would crewing answer quickly enough to avoid having to listen to the one noise that was worse than the static on HF? Who needs waterboarding when the world has muzak? It was a shame that the company would go to all the trouble of recording every conversation, yet, would not go to the trouble of licensing a soundtrack that does not have the same effect on a person's nervous system as dragging nails down a blackboard.

Fortunately, crewing was on the ball, and I was spared the jazz sax horror. As soon as I told the young chap on the phone who I was, but before I could say to him why I was calling, he cut in nervously.

'Oh, hi. Yes, err, err, I have got to put you through to the duty pilot.'

Considering this manager did not work in crewing, I was a little perplexed why he would need to talk to me. Maybe he was going to thank me on behalf of the airline for saving them around £250,000 and for letting them know about my unavailability tomorrow?

The duty management pilot that evening was Andy Thorington, who was Group Head of Airbus Operations (GHAO). His role involved safety and regulatory compliance. He would, or at least should, know the score.

It was immediately apparent that this was not going to be a one-on-one private conversation. The several people eavesdropping disobeyed the very first rule about listening covertly: don't collectively whisper *'yeah shhh he's on now shhh'*. GCHQ, it was not. More like an episode of Dad's Army.

Eventually, there was a voice that I was supposed to hear: 'Hello? Mike.'

'Andrew, I am supposed to be speaking to crewing; I am not sure why I have been passed through to you.'

Thorington explained it was because I was refusing to fly tomorrow. Strange that, without knowing any of the details of why I would not be operating, Thorington, with the assembled Inspector Clouseau incompetents as an audience was quizzing me about it. Pity they were not so on the ball sorting out two pilots in ten hours or preventing this whole cock-up in the first place. Perhaps they had an ulterior motive?

I explained to Thorington that I had spoken to the operations department on HF and asked them to pass on the message to crewing that I would not be available for tomorrow's flight. In easy to understand terms, I advised him of my suspicion that during tomorrow's duty, I would become so tired it would constitute fatigue. In simple terms, it would be unsafe. I also explained that the company had rostered my next duty just over the minimum legal rest period. I was simply giving, as I was required to do, an early heads up to crewing that they would need to find another pilot to replace me.

'There is no point in doing that as we don't have the crew anyway due to all the training and simulator duties' was Thorington's less than helpful response. I explained that if we did or we didn't have enough crew was way above my pay grade. I was merely giving the company as much notice as possible.

In complete contradiction to that which the operations manual explicitly stated, Thorington pronounced:

'It's premature and inappropriate to determine now that you're not going to be fit for the duty when you have a good rest period.'

Woah, hold on there cowboy! Thorington was informing a senior captain that it was '*inappropriate*'? Who the hell did he think he was? Had he any medical training? What exactly was this '*good rest period*' Thorington was talking about? Did he mean the rest period that was only just over the absolute legal minimum rest period, after what would be, by the time I got home, an 18-hour working day? My attitude instantly hardened.

This implied pressure was absolutely not acceptable. If management were doing it to me, you could be sure they were doing it to other crew. Given what Thorington had just said, it was clear he was willing to promote ignoring the companies laid down procedures if it conflicted with commercial considerations? Was this why Thorington had been promoted to the position he was in?

If he thought for a moment that he could influence what I considered to be the right decision, then he was in for a shock, as were all those who were listening in. The management style now adopted by the company since Hutchings joined was one of harassment and bullying. Sorry pal, wrong guy to try and bully and the wrong guy to try and harass. Time to let Thorington know I had his number.

'By the way, this call is company business; therefore, it is still duty time,' I informed him.

The subtext was, keep talking laddie, my duty period is still running and eating into the rest period, making it even more unlikely that I will change my mind. I did consider saying something a little more colourful, but this would be sufficient to get my point across.

Immediately Thorington got the message that what worked on the rest of the pilot community had backfired. As most bullies do when confronted, he backed off.

'I understand that it could have been a demanding set of duties. Go home, get a good night's sleep and if you feel you're not fit for tomorrow's duty, ring tomorrow morning.'

I agreed I would do this but also reminded him how much money I had just saved the company and I had not even had a thank you. A '*thank you*' was then proffered that held as much sincerity as a mid-western televangelist. I hung up. All that remained to do now was to drive the one hour fifteen minutes home. The adrenaline pumping through my system from the recent phone call got me around halfway before the fight or flight chemical began to dissipate. I became so weary that I began to worry I might fall asleep at the wheel. Turning up the Rammstein CD currently attempting to turn my speakers inside out had no effect. I admitted defeat. I pulled over to allow the gathering fog to envelope my consciousness for a few golden moments. I reclined the seat and closed my eyes. BAM! I awoke with a start. What should have been the first bit of quiet since 04:30z that morning would, every few minutes, be shattered. I would fall asleep for a few moments then wake with a start. Was I still driving? Had I fallen asleep while careering down the road at 70 mph? I didn't know anymore.

I eventually walked through my front door at 22:30z (half past eleven at night). I did not even have anything to eat, not even a glass of wine; all I wanted was my bed. I arrived home only three hours before I had got up for two of the three days previously and only five and a half hours before I had woken up that morning. Eighteen bloody hours and I was spent.

Those with little knowledge of sleep science would assume if somebody is extremely tired they would fall asleep for as long as needed and would wake up refreshed. The human body, however, simply does not work like that. Those of you who have flown from the UK to Las Vegas or San Francisco will have experienced this phenomenon. West coast USA is eight hours behind UK time. Imagine when you arrive you decide to force your body to shift to local time, and you manage to stay awake until 22:00 Vegas time. Your body clock, however, is still on UK time; therefore, as far as your vital organs are concerned, it is 06:00 the following morning. If you had flown the aircraft that would mean you would have been awake for around twenty-four hours. You would be, by now, totally spent and desperate for sleep. You would imagine you would now get at least eight hours' sleep. Nope. Try it, I have, about thirty times. I can pretty much guarantee you that if you go to sleep at 22:00, within three hours or so, you will be wide awake because your circadian rhythm/ body clock is still on UK time where it is now around 09:00. This is precisely what causes jet lag, or to give it its correct name, circadian disruption. The same effect is evident for people who work shift patterns. Now consider the time differences at different destinations. Now mix in the added complexity of operating short-haul flights when workdays shift from very late nights to very early mornings and back again. Given the above, is it any surprise why the average life expectancy of a pilot who retires at sixty-five is sixty-seven?

My run of duties over the last few days had started on Friday 2nd May. I had got home at 21:10z. I then had the medical situation to deal with. This followed by two early morning wake-ups at 02:00z on the 4th and 5th May and a 04:30z wake this morning. I eventually got home after being out of the house for eighteen hours. It was a racing certainty my circadian had had its rhythm drummed out of it this week. As I could have predicted, by 04:30z, I was wide-awake. No matter how hard I tried, Mr Sandman had left the building and was not coming back any time soon.

By 08:30z, I had given up any hope of getting back to sleep even though I still felt utterly drained. I now had to ask myself a straightforward question. Would my level of tiredness constitute a risk to today's flight if I reported for the next flying duty at 11:50z? It would mean driving back to work in just over an hour. It was one of the easiest decisions of my aviation career. At least it should have been. I suspected from the way Thorington had attempted to lecture me the previous night, combined with the fact that the conversation had an audience, the brown stuff would be hitting the rotating metal blades. Well, you know what? Let it hit! Fuck the bloody lot of them. I had never committed a criminal act in my life. I was not about to start now for this bloody ramshackle outfit.

But hold on, maybe I was being far too hasty. Maybe I was jumping to conclusions that were not actually there. I had not been threatened with any repercussions if I refused the duty today. In fact, Thorington himself had said to go home and get some rest and if I still believed I was not fit, to phone crewing and tell them. Yes, maybe I was wrong in my suspicions. One of the signs of fatigue is a short fuse. This morning mine was virtually non-existent.

I calculated if I agreed to fly today's duty, it would mean landing a Boeing 767 back into one of the busiest airports in the UK, at night, with nearly three hundred passengers on board plus crew after being awake for nineteen hours. This was even before any possible delays were factored in.

It was a no-brainer. I was already in no fit state to fly, so God knows how I would feel on the return from Turkey. The duty was scheduled to land around midnight. It just did not bear thinking about.

Throughout my entire flying career, to ensure I always make the right decision, I would ask myself the fundamental question: 'Would I be happy to have my family as passengers on board this aircraft?' This question must have some legitimacy in that throughout my entire career, not one of my command decisions has ever resulted in a '*reduced safety margin*'.

For the first time in my entire flying career, I was going to phone an airline and tell them I was refusing a duty due to fatigue. There was another way out of this predicament. I could ring the company and lie. I could tell them I was ill, tell them I had a head cold or my ears were blocked. The result would be precisely the same. I would preserve that safety margin with no chance of any repercussions. If I reported fatigued, then questions would be asked, fatigue reports filed, and a document trail created.

Was I going to allow the implied pressure from Thomas Cook to affect a decision that only I could make? From the very first flying lesson in 1989 with the Californian beach babe, right through to now, I had never allowed anybody to pressurise me into doing something that I knew was either illegal or unsafe. As I lay there looking up at the ceiling in my bedroom, these thoughts swirled around my head until the dust cleared and the decision became crystal clear.

Q. Were the company short of pilots?

A. Yes.

Q. Should I allow that fact to *infect* my decision?

A. Absolutely not.

Q. Was I ill?

A. No.

Q. Was I so tired it would impair my alertness and ability to operate an aircraft

safely?

A. Absolutely yes. That was the ICAO definition of being fatigued.

Q. Is it a criminal act to fly if you know or suspect you are or will be fatigued?

A. Yes.

It then hit me like a ton of bricks. If with all my experience I was feeling an implied pressure to fly, or at the very least lie about why I was refusing to fly, how would a young first officer react to the same situation?

I had always done 'the right thing' in my flying career, and I knew if I didn't do the right thing today, totally irrespective of if I got away with it, I would always know inside that, when the crunch came, I buckled. I placed my career before safety.

All this time, money and effort to become a professional airline pilot would stand for nought if I allowed myself to be coerced into performing a criminal, unsafe act.

But, what if I was making a mountain out of a molehill? What if I was imposing my own cynical view brought about by yesterday's antics by the company, and the phone snooping during my call with Thorington? Maybe yesterday's calamity was an aberration. Whatever it was, this morning, I had drawn my own line in the sand. After all, it was only mine to draw.

EU-OPS 1.1090 4.1

"A crew member shall not operate an aeroplane if he/she knows that he/ she is suffering from or is likely to suffer from fatigue or feels unfit to the extent the flight may be endangered."

I picked up the phone and rang Thomas Cook crewing.

*'Thank you for contacting the Thomas Cook operations and crewing information system. Please be aware all calls will be recorded for training and quality purposes.'* Muzak, they must be busy. Eventually, a crewing controller answered.

I began 'Just to let you know and to confirm, as per my discussion with Andy Thorington last night, I will not be available for today's duty due to fatigue. I shall fill in a fatigue report form later.' The crewing controller seemed a little taken aback.

I reassured him:

'Don't worry about anything. I will email the chief pilot to let him know the details.'

With that, bleary-eyed, I set about writing an email to the ex BALPA CC chairman, Scadeng. As I wrote the email, it suddenly dawned on me. Only a week had gone by since Scadeng had phoned me up telling me to change a witness statement. Was that what was behind what had happened yesterday? Was this an attempt by Scadeng and Hutchings, and by the sound of it Thorington, to engineer a situation for refusing to change the witness statement? This all seemed a little too much of a coincidence. What if I had bowed to pressure and operated? Were they really willing to place passengers in danger?

I began the email to the new incumbent.

The email in full sent the morning of May 7th to Roger Scadeng:

"Dear Roger,

Apologies in advance for needing your input on this, I am sure in your new role you have more than enough on your plate than dealing with issues such as this, but I feel that it is important that you are made fully aware of what is happening 'on the line.' I am writing to you directly as I did not feel that the questions that the situation raises would be appropriate on Xplorer.[Thomas Cook's own forum]

For the first time in my flying career, today I rang up crewing to inform them that I would be unable to perform a flight duty, not due to illness, but due to my very real concern that I was experiencing the onset of fatigue which would erode flight safety if I completed today's duty.

As you will see if you read my comments on my flight report of yesterday 6th May, there are some serious issues that need to be attended to regarding the realistic rostering of flights and the 'back-up' that should be afforded to flight crew for the day to day operation of flights.

I rang crewing upon landing last night and was put through to the duty pilot, Capt. Thorington. He kindly explained the intricacies of captain's discretion, and that in 'his' opinion I could not state with any certainty that I would not be fit for today's duty in advance; he informed me that due to sim training etc etc etc crewing did not have the luxury of spare crew to re- crew the flight last night. On that basis, I feel the full details of the situation need to be known and for your view on the matter. Although I do not have information in regards to if we do, or do not, have enough crew, (way beyond my pay scale), the single reason I rang crewing to inform them last night, of the high probability of not being fit for today's duty was to allow the company plenty of time to arrange cover. As it was, it was not until around 9:30, that I rang crewing today having established I would not be fit to fly a long duty day today. I am in the process of filling in a fatigue form.

My run of duties began with a standby from 06:00 on Fri 2nd May. I was called out to position by road to travel to GLA and then immediately fly an aircraft back to Manchester. I was off duty at 19:45 Zulu.

My next duty was supposed to be a 4:45 Z report PMI. This was obviously changed due to the previous duty to a rest day, (although this day was used to prepare for the sim sessions.) On Sunday 4th and Monday 5th May I had two sim sessions, both starting at 04:15Z On 6th May, I was originally on standby, however this was also changed to a flight. The duty was as follows. Report 0700Z to fly MAN-HER-AYT-MAN.

I had a look at the flight times after my sim session and realised that the duty time would amount to 12:30. The allowable FDP was 12:30. It is important to note that the flight was rostered with a 0800 LOCAL report. Had it been rostered one minute earlier, the allowable duty time would have been 11:30, i.e. the flight would have been rostered one hour into discretion.

I rang crewing to query this and suggested, due to my previous duties and that there appeared to be absolutely no 'fat' in duty times that it would be advisable to have another crew available to operate the aircraft home. I was told no crew were available. I then suggested that a crew member who had not previously reported for two duties at 04:15Z, and therefore less likely to become fatigued, be called out for the duty. Again I was told no one was available.

Upon reporting yesterday, the reality of the situation became apparent. Upon looking at the initial plogs [flight plans] produced, at standard cruise and turnaround times, it became clear that the duty was actually planned into discretion by approximately twenty minutes. I informed the company that I would not be utilising my captain's discretion. As per FTL 7.18.1, it states 'The extension shall be calculated according to what actually happens, not on what was planned to happen.' Therefore, as it was planned to go into discretion, my captain's discretion would not be an option. The company then produced high speed cruise plogs (Mach .82) in an attempt to reduce the planned FOP to below that of 12:30. Even then the duty time showed in excess of the maximum of 12:30.

If Mach .82 was not sufficient to allow the series of flights to be operated within the legal FDP, what figures were used to come up with the original FDP of exactly 12:30 which, as I have already mentioned, was exactly the maximum FDP available?

The only reason I elected to continue with the duty once I had seen the times that were 'planned' on the flight plan stage was that I was assured by the company that if I was not able to make up time on turnarounds, then another crew would be dispatched to continue the flight should I need to divert due to duty hours. As as per 7.18.1, I was not able to consider discretion because the flight had been planned to go into discretion.

Needless to say, on the flight back to Manchester it was clear that despite cruising at Mach .82 and requesting direct routings all the way from ATC we were still going to be fifteen or so minutes into discretion. As agreed I called up the company via HF to inform them that we would have to divert due to the FTL [Flight time limitations] working on the information that I had previously been given. I was then told there was absolutely no crew available to fly onwards to MAN if I diverted and that the aircraft (and of course paying passengers) would be grounded.

Having now been completely let down by the company on this matter, having previously being told that another crew would be available to take the flight onwards, I had to make a decision based on this new 'information.' Taking into account the ability of the senior first officer, and the fact that he had only just started his series of duties, I elected on this occasion to exercise captain's discretion to get the company out of the massive financial hole that was due entirely to its own making.

Upon landing I was very tired, and more to the point extremely disappointed in the company's handling of the matter. After the phone call with Capt. Thorington and completing the voyage report, I rang crewing at 20:45 and they showed me off duty at 21:00. After

getting home and going straight to bed, I awoke at 5:30 and was unable to get back to sleep, no doubt due to the three previous early morning waking times. By 9:30 I was confident that I would not be fit due to the probable onset of fatigue to complete today's flight safely, therefore I contacted crewing.

Based on those above timings today's flight would have been within 55 mins of minimum rest.

Hopefully you will agree that my series of duties were absolutely not ideal and although legal, (I shall leave the FDP of yesterday's fiasco out of that comment for the time being), would have resulted in me operating a flight back this evening at a reduced level of alertness that would have been prejudicial to the level of safety that I would expect.

For the record I have never, when the flight has been legally rostered, refused to utilise my legal command discretion for the company's benefit, and as yesterday, my willingness to do so has saved the company many thousands of pounds.

As you are also aware, in regards to the other matter that you rang me on 30th April, I still have not received even the courtesy of a reply from the solicitors, and no request to 'sign' the witness statement. Again, I do wish to draw your attention that on two of my recent days off I have been contacted and requested to be involved in 'company business' on this matter. I have not asked for those days off back in lieu or for day-off payments. I think you are fully aware of my feelings of not only the solicitor's handling of the case, but also the complete disassociation and lack of input from the company on the matter over nearly four years.

I am sure and I hope you will take the details in this email in the manner that they were intended.

Regards

Mike Simkins"

I was actually, factually, incorrect when I said in my email to the Chief Pilot that '*my series of duties were absolutely not ideal and although legal'*.

Mistakes do occur when fatigued, after all. In actuality, the series of duties planned by Thomas Cook were very much not legal.

EU Ops state:

"Operators shall ensure flight duty periods are planned to enable crew to remain sufficiently free from fatigue so they can operate to a satisfactory level of safety under all circumstances."

ARTICLE 145 (2) OF THE AIR NAVIGATION ORDER

"The operator of an aircraft to which this article applies must not cause or permit any person to fly as a member of its crew who the operator knows or has reason to believe is suffering from or, having regard to the circumstances of the flight to be undertaken, is likely to suffer from, such fatigue as may endanger the safety of the aircraft or of its occupants."

I had informed the airline two days previously that the run of duties were fatiguing. I was fatigued. Therefore, Thomas Cook had committed a criminal act in planning those duties.

After I sent the email, I set about filling in a fatigue report that was sent to the safety department of the airline.

There were a lot of legal reasons why what I was being asked to do was wrong. It was, in fact, given my understanding of the law, criminal, but let's get down to basics and consider this little scenario.

You have had a lovely time with your family on a beach holiday in Turkey. The last day of the holiday comes around. You and the family make the most of the all-inclusive buffet before the coach arrives to drive your family to the airport. The postcards have been sent, and you still have a little local currency left to get a couple of drinks for the children and beers for yourself and wife at the airport. Checked in, security negotiated, you finally board the aircraft and make yourself comfortable. You decide that you are going to watch the in-flight movie, the lights go down, the children fall asleep, four hours of peace before you land back into Manchester just before midnight. Are you feeling relaxed?

Now consider this. Your captain flying you home has had the same run of duties that I had. Unlike myself, he had allowed himself to be pressurised into operating the duty and was sitting at the controls of the aircraft you and your family are currently aboard. Given this information, are you still feeling relaxed? Would you allow your family to board the aircraft if you knew how tired your captain was? How confident would you feel about the decisions the man with four bars would be making? Consider how you would feel if you knew that, at the most dangerous part of any flight, the approach and landing, he had been awake for 19 hours? I know I wouldn't trust his abilities. I certainly would not have allowed my family to board that aircraft.

That is why I refused to fly the aircraft to Izmir on May 7th. It was such a shame that an Air Canada captain did not make the same call in 2017, but more of that in a later volume.

As the day progressed, it became more and more apparent I had made absolutely the right decision, dozing off several times that afternoon, including, for one period, a couple of hours.

An airline management team that placed safety before profit and had an open and transparent fatigue reporting system would not only accept any crew member's decision to call in fatigued, they would embrace it as someone doing 'the right thing'.

I would find out over the next couple of days what the culture within the airline I worked for was.

At midday, on May 9$^{th}$, the phone rang.

# CHAPTER FIFTEEN

## *A Bit Of Gardening*

*"Employers may record their employees in secret only in very exceptional circumstances. This is where it is suspected that criminal activity has taken place."*

Information Commissioner Office

Lisa, brought me the phone. It was one of the new, never-flown-an-aircraft-in-their-life pilot managers (PM). Out of respect, I will withhold her name as very sadly, she has since passed away. I shall refer to her as 'Julie'. The fact that Hutchings had removed all the experienced pilots from the role of pilot managers and replaced them with individuals with no experience of the rules and regulations that a pilot must comply with undoubtedly was a significant issue underlying the following conversation.

The following is an exact transcript of the phone call. It was 9th May at 10:19 to be precise:

PM: Hello Mike, how are you okay to talk?

ME: Yes

PM: I need to discuss a serious allegation

ME: Ok

PM: It has been alleged that you refused to operate a rostered duty. This is potentially gross misconduct. Therefore, due to the seriousness of this allegation I am suspending you from duty pending investigation as under the airline's disciplinary policy. This is with immediate effect and will be on full pay. I want to be clear that this is not a prejudgement of the outcome of the investigation nor is it disciplinary action. During the period of your suspension you must not attend any airline premises unless asked to do so by myself, HR or the another pilot manager and you must not attempt to contact any member of staff with the exception of HR, the other pilot manager or myself or union representative. The nature of your suspension must remain confidential. We will endeavour to ensure that the investigation is completed in a timely fashion; we would hope to be in contact regarding the outcome of the investigation within two weeks where possible. You will be asked to attend an investigation meeting in the next few days; we will write to you to invite you to this meeting also to formalise your suspension. I understand that this will be a difficult time for you and would like to offer you any support we can. There is an employee assistance program who can be contacted, please give them a call if you'd like to discuss this confidentially with someone.

Your pass will be temporarily suspended. If you need to speak to anyone please call me any time today and the other pilot manager after the weekend as I'm away.

ME: I will not be contacting the company, I will go through BALPA or the CAA, goodbye.

I was never one for long, drawn-out flowery speeches, but my view at this point was as soon as BALPA heard about what had gone on there would be uproar. I had nothing more to say and nothing but contempt.

I was right to question management integrity. It was clear now it was not just a fatigued ex-drummers cynicism.

The reason I can recite the conversation word for word was that the airline recorded it. As you see from the transcript, at no point within that conversation was I informed that it was being recorded?

I was to discover, from the Information Commissioner no less, that in an employment context, the data protection code, says that:

"Employers may record their employees in secret only in very exceptional circumstances. This is where it is suspected that criminal activity has taken place." They also state:

"The covert monitoring of workers can rarely be justified. Do not carry it out unless it has been authorised at the highest level in your business."

Given that the highest level was Hutchings and Scadeng, it would be reasonable to assume at least one of them must have sanctioned the recording.

My suspicions with regards to the company's motives were beginning to coalesce. At no point in my entire aviation career had I ever been suspended or had any disciplinary issues whatsoever. So why now?

Was this an attempt by the new chief pilot to put a marker down to impress those higher up the food chain of his management style? With so few pilots now left to cover the flying programme, more and more pilots were going to have to work harder to cover it. Was this a none-too-subtle message to warn off other crew reporting fatigued?

The unfortunate thing was, again, this call came as no shock. With Hutchings now installed as director of flight operations, the acceptance by the ex-union chairman to, as far as I could conclude, become his lapdog and the removal of any pilot managers who were actually pilots who would question management motives, the culture shift within the airline was seismic, and, in my opinion, it was dangerous.

It was the beginning of May, the start of the busiest period for the airline. This was an airline that the ops team had told me two days ago were not 'flush with crew'. Yet, here they were, suspending a senior captain with over 13,000 hours of flight time for refusing a duty. One that, in his own considered opinion, based on an in-depth knowledge of aviation law and the understanding of his own diminished abilities was caused directly by a criminally- planned, fatiguing series of duties.

I was still trying to make sense of what had happened. Could this be put down to some administrative cock-up? Surely, common sense would dictate the last thing any airline would want is pilots flying at a 'reduced safety margins'. That is, unless of course, those involved were devoid of any concept of what the PR fall-out of that attitude would be.

Was I concerned? The answer was no.

If this did turn out to be anything other than an admin cock-up or misunderstanding, it would say more about the mentality of the senior Thomas Cook management team than it would about my attitude. This was an attitude that had flown passengers safely for the company for the last sixteen years, for fourteen of those years as captain. The attitude was one that had never been involved in any serious incident, never refused a duty due to fatigue before 7th May and had an exemplary sickness record.

Yep. I was a regular little goody–two-shoes.

A few days later, I received my confirmation of suspension letter, which added a further restriction. I was now banned from carrying out any work on behalf of the airline or accessing any company files or electronic mail. It also informed me that Jo Burke was going to handle the case. I strongly suspected that the pilot manager who had suspended me in the first place had since realised this was not a situation she would want to get involved in. Burke was left holding the baby.

According to the phone call, it looked like I would be having around two weeks' paid leave. Worst things have happened at sea. The sun was out, the grass needed cutting, so I better make the most of this unexpected holiday. Once it was explained to Burke the minutia of legislation relating to fatigue, I strongly suspected I would be getting an apology. Possibly a shake of the hand, and probably a beer bought for me at the Romper pub as a peace offering. At this stage, that was all that would have been required.

Just in case an apology was not forthcoming, I rang a member of the BALPA CC and let them know what had happened. I probably would not need BALPA help but, better safe than sorry.

I spoke to Richard, one of the BALPA CC reps. He had been on the CC with Scadeng for a few years. I explained what had happened.

'Come on, mate, I find that hard to believe. There must be more to it.'

Richard was having difficulty getting his head around the fact that the company were suspending a senior captain for refusing to fly due to fatigue. After the call, he asked me to put the details in an email to him. His response was enlightening.

'There is supposed to be an open reporting culture as far as fatigue is concerned, and the CAA are worried that there is under-reporting going on because pilots are afraid to report it.'

He summed up the current situation very eloquently: 'It's fucking barmy!'

Richard then mentioned that there was a meeting of the BALPA CC in a few days, and he would raise the issue there. Within a few days, he rang back.

'I've got a feeling somebody's been briefing against you, the BALPA CC does not support your position.'

I was floored. For years BALPA had claimed the number one concern of its pilot membership was that of fatigue. Here was a blatant case of a pilot being suspended for informing the airline, as he is legally required to do, that it would be unsafe to fly due to fatigue, yet they did not support me!

Some of the more cynical of you might conclude this was due to Scadeng, when he was union chairman, building a little BALPA clique around him. A number of my colleagues had suggested this was indeed the case. Although they might say that, I, however, could not possibly comment.

Richard suggested: 'Give Ian White a call at BAPLA's head office; he is the expert in flight time limitations for the union.'

Thanking Richard for the heads up, I was speaking to Ian moments later.

Ian had precisely the same reservations as did the rep.

'I'm sorry, but I'm finding it hard to believe that a company of this size would do something like this; it goes against all common sense. There must be more to it, are you telling me everything?'

Again I explained the sequence of events, and why I refused to operate the flight on 7th May. Ian agreed that with the run of duties I had done he would be amazed if anyone would be safe to drive a car, never mind an aircraft.

I was struggling to comprehend that an airline would be so singularly stupid to go down this avenue. It just made no sense at all unless, as I had already suspected, it was a cynical attempt to intimidate line pilots not to report that they were fatigued.

I had explained to both Ian and Richard that despite explicitly asking for his view and involvement, I still had no reply to the email I had sent Scadeng.

Was this an attempt by Thomas Cook to bully and harass me? As a schoolboy, it had taken me way too long to work out how to deal with schoolyard thugs. My eventual solution was to wait until one of them was on their own and give them the biggest beating they had ever had. Always away from prying eyes, of course. Word soon got around, and it seemed to solve the problem. It gave me a taste for confronting bullies that has lasted till this day. If it was a case of bullying, I was still hopeful that, like the schoolyard bullies, they would soon realise they would be onto a hiding to nothing.

My next week's roster had been wiped of duties and replaced by a roster code indicating I was suspended. The following three weeks of duties, however, were still there. Maybe I would be back online next week after a handshake and an apology?

After being suspended for a week, I had an email from Jo Burke:

"We are currently completing our investigation and have several people to meet with, therefore due to shifts and people on leave, it may take a few weeks. If you need anything in the meantime, you have my details to get in touch."

So it had changed from "about two weeks" to "a few weeks". Why exactly was Thomas Cook going to such lengths when Scadeng could easily sort it by replying to my email? Maybe he saw the email as *'entrapment'* in that it removed any chance for him to claim plausible deniability? Perhaps he was right. Maybe that was precisely why I sent it.

Only a few days later I had a phone call from Lindsey McCulloch who was Thomas Cook's new employment lawyer. They were still banging on about the bloody witness statement I had refused to change. She asked me again to alter it despite my already telling Scadeng the reasons that I wouldn't.

Time for a bit of fun.

'Sorry, I'm suspended, and I can't speak to any Thomas Cook employees apart from HR and the pilot managers, and you are neither.'

I then had several minutes where McCulloch gave me every reason under the sun why I would not be breaching my suspension conditions by speaking to her. Naturally, I declined, informing her if she wanted to talk to me she would have to get a pilot manager to authorise it.

It was amazing how quickly Thomas Cook could react when they wanted something. Within a couple of hours, I had an email from Jo Burke informing me that I could now speak to McCulloch.

McCulloch duly rang up and again gave me the soft soap why I should agree to alter my witness statement. Again I politely declined.

Within a few hours, my roster was wiped. All the flights for that month, at a stroke, had been removed. Of course, my suspension for refusing to fly when fatigued had nothing whatsoever to do with me refusing to alter a witness statement.

An unforgivable lack of crew with millions being spent on bringing in outside airlines to cover the fact that Thomas Cook did not have enough pilots to cover the duties, and yet, the investigation was it was going to 'take a few weeks'. Nope, in no way was this suspicious at all.

# CHAPTER SIXTEEN

## *Take The Bait*

*"If accidents are only the visible tip of the iceberg, then pilot fatigue is a major cause for concern as a potent risk factor in accidents waiting to happen."*

Dr Hudson (Thomas Cook's Medical Advisor)

By now the garden was looking fantastic, and we were into June. The 'unforgivable lack of crew' at Thomas Cook was now even making the news headlines, with the airline having to cancel flights left, right, and centre. On the second of June, it was reported that hundreds of passengers were left stranded for eleven hours at Bristol Airport after two Thomas Cook flights were delayed.

The excuse?

*'We'd like to unreservedly apologise to our customers for the significant delay to their flight from Bristol, which has been caused by the impact of short term availability of crew, including sickness.'*

Funnily enough, there was no mention that the lack of availability of crews was due to the complete and utter train-wreck Hutchings and his chums had caused by slashing pilot numbers. At this point, of course, Hutchings was still trying to convince anyone who would still listen to him that we had the 'correct establishment figures'. Maybe the Thomas Cooks PR department had come to the reasonable conclusion that the real reason passengers who had paid thousands of pounds were having their holidays ruined would not look to good in print? No, pilot 'sickness'—that sounds a lot better.

If pilots were working their days off, and working longer hours driven by lack of crew, it would be fair to expect the number of fatigue reports to increase significantly. Bizarrely, what *was* happening over this period was that there was a reduction in the number of fatigue reports. There was, however, a doubling of days off due to sickness. It was not rocket science to determine why. The company had already displayed what would happen to anyone who refused to fly due to fatigue.

On the other hand, if a pilot fears retribution by the company for telling the truth and reports sick instead, then no investigation is needed. A lot less hassle, a lot more straightforward. There was, however, a downside to hiding the real issue for those that reported sick.

The cabin crew had recently found this out the hard way. When it came to redundancies, Thomas Cook intended to implement a process known as the Bradford scale. It meant they were basing layoffs, not exclusively on seniority but on how many sick days an individual had taken off.

It was a posting I had made on the Thomas Cook-hosted forum concerning the Bradford scale that was probably the first time I ever appeared as a blip on Hutchings' radar. It concerned two separate but connected conversations I had with cabin crew on two distinct occasions.

The first conversation occurred when I arrived at the aircraft at the start of a duty. As I entered the galley, there was a young member of cabin crew distraught. She was sobbing uncontrollably.

I asked what the problem was. She told me she had just been reprimanded by one of the base managers because her 'hair bobble' was the wrong colour.

I was just about to put my foot in it and advise her not to take things so seriously when she added:

'And I already felt unwell before I got here', followed by 'and they have not even paid me for last month yet'. It transpired Thomas Cook had failed to pay its cabin crew staff for several weeks. When the crew had complained, highlighting they could not afford the petrol to get to work, the company had sent taxis to pick them up.

'If they have not paid you, why are you in work? You shouldn't be flying if you are ill anyway?'

Her response showed how these, predominantly young girls, were treated.

'I'm on a 7/5 contract, and if I make trouble, they won't ask me back next year.'

A vast proportion of the Thomas Cook cabin crew worked seven months over the spring and summer and were then laid off until the next year. It benefited Thomas Cook as they were able to avoid having to comply with specific rights for full-time employees. It also meant they held the sword of Damocles over anyone who complained about not getting paid or wore the wrong shade of bobble.

I explained that having seen the state she was in, in my capacity as the captain, I was sending her home. She was not fit to fly. I explained if she had any comeback at all from the company, she should let me know.

I heard nothing more.

The next conversation was a week or so later with a very experienced senior cabin crew member. We were in the cruise when she brought in drinks and sat down on the jump seat. I could tell she wanted to talk. She was seething. She informed me that the company were making redundancies, but not in seniority order. Instead, they were getting rid of crew based on how many 'Bradford points' they had amassed. She had been off work for three weeks following a serious operation. Those weeks meant, due to her 'points', the fifteen years at the company counted for little. She was going to be made redundant. It was the first time I had ever heard anything about this palpably unfair situation.

As a result, I posted a question to the management on the Thomas Cook internal forum.

I pointed out that crew being made redundant based on sick days was potentially a serious safety issue. Did the company really want crew turning up for work when they knew they were unfit to do so, just to avoid collecting points?

I also referred to the first young girl I had sent home and raised a valid point. Had it got to the stage where a captain had to intervene and stand down crew because the attitude of the management resulted in crews being terrified to not report for work? I added that although the Bradford scale might be appropriate in other industries, according to the law if a crew member reported for work and operated a duty when they knew they were unfit, they had committed a criminal act. I concluded that, if this were the case, would this not leave the company open to any legal action if they made crew redundant based on the Bradford scale.

Hutchings replied that he had:

*'every confidence'* that the director responsible for cabin crew, Maggie Kennedy, would deal with the situation *'appropriately'*, and that the Bradford scale was only *'part of the process'*.

Hutchings had already set a precedent in ignoring seniority with the recently-enforced base moves. So why not scrap it altogether in favour of the Bradford scale for pilots as well? Given BALPA's toothless efforts over the past couple of years, they were hardly going to put up much of a fight.

I don't think I had ever been so popular with colleagues, apart from when it was my shout in the bar down route. I would regularly get emails and phone calls informing me of what was going on. One colleague rang me to tell me he had fallen asleep on a flight, only to wake up and find the first officer was also sleeping. He also told me he was afraid to refuse fatiguing duties based on what had happened to me. It was by no means unusual. It seemed to be becoming the norm. It was textbook normalisation of deviance.

Some of my colleagues would blind copy me in on emails they were sending to Thomas Cook's head of crewing, the Fatigue Risk Monitoring System Manager (FRMS) Jane Marsden.

Example of what I was sent included;

"I cannot think I have ever flown such a fatiguing roster. In this matter, I can speak for many of my colleagues on the line."

And:

"I am just as concerned as you, that our pilots are not reporting enough and that our flying regime would perhaps explain why sickness rates have doubled."

And:

"our management are adopting quite intimidatory practices."

My situation was also referred to;

"I am sure you are aware of the rumour that a pilot has been suspended from duty for calling fatigued and this has spread like wildfire and would not, I suggest have helped FRMS, perhaps exacerbating the above problems."

Another example of how bad the situation had got under Hutchings and Scadeng watch was:

"A first officer told me that in the cruise, his captain fell asleep twice whilst at the controls without warning on the return sector of a bullet. There was no embarrassment on the skipper's behalf, just a dangerous fatigue level. Numerous more stories to tell, all unreported."

It looked like the airline's cunning plan was bearing fruit. The evidence appeared to show that Thomas Cook pilots would prefer to fly at a *'reduced safety margin'* and fall asleep at the controls rather than refuse the flight. It appeared Thomas Cook were still committing the illegal act of planning duties that *did* result in fatigue.

The only logical conclusion was that pilots had allowed themselves to become intimidated by the bully, terrified of the consequences should they refuse a duty. Could anyone respect such a feeble group of individuals? Thomas Cook management certainly didn't appear to.

I still had no reply from my email to Scadeng, but finally, I was invited to an investigative meeting. This was over a month after I was told it should take two weeks for the entire investigation.

Richard the BALPA rep and Ian White from head office agreed to accompany me to the meeting. No one else on either the BALPA CC or head office was interested in the case. Considering I had received no help or advice from either. I was, therefore, grateful for Richard and Ian taking the time out to support me.

I met them both in a hotel by the airport a couple of hours before the meeting. Ian and Richard were in total agreement. They both believed that Burke was going to unveil, within the investigation meeting, a hitherto undisclosed misdemeanour so heinous that I should be thankful I was not already chained in irons.

Ian remarked that if what I had told him was right, he had never seen such abuse by any airline of a captain for merely doing his job. He sounded almost apologetic when he said there has got to be something else going on or they have something else on you. The situation seemed so ridiculous even I started to wonder had I done something wrong. Something that I was completely unaware of that would warrant such a reaction. That was, of course, unless a refusal to alter a witness statement justified a suspension.

We eventually arrived in reception at the Thomas Cook offices. Several pilot colleagues walked past as we waited. The considered consensus was to tell the company to go forth and procreate and then go away. It was a shame my colleagues were not as vocal directly to Hutchings and Scadeng. If they had, then things might not have descended to this level.

Some ten minutes after the appointed time Burke and Karen Harragan, a member of HR arrived.

'I'm sorry, Mike, who is this?' asked Burke. My little surprise of bringing Ian along had worked very nicely. Previously, Burke was at pains to explicitly state that the investigative meeting was entirely a fact-finding exercise. Considering Ian was a flight time limitations expert, and she wasn't, I explained Ian could help in her 'fact-finding'. It did not seem to go down too well with Burke. I didn't for one moment, think it would. That's precisely why I asked him to attend.

It was apparent to Ian, Richard and myself; she was already out of her depth, and this was before the meeting had even commenced. Her demeanour reminded me of a child in the swimming pool who had just stepped into the deep end and was frantically trying to get back, so her feet could touch the tiled floor again.

'Erm hold on, I'll have to discuss this.' With that, both she and Harragan scuttled back up the staircase they had just descended. Now, who possibly did they discuss the bonus of having an expert in the room with? If it were Scadeng, he would probably have enough spare time on his hands. He certainly was not busy replying to my email.

Seconds out! They were back—well actually about 10 minutes later they were back.

Burke delivered the message. Ian was not welcome at the meeting.

'But if you want an expert to be involved in your fact-finding mission, surely he needs to be in at some point?' I smashed the ball firmly back into her court.

Burke was flustered. A red patch 'tell' appeared on her neck giving the game away that she had become uncomfortable. A future playing poker would not be a good career move.

'Erm err ok, well maybe towards the end of the meeting.'

That was good enough for me.

Richard and I followed Burke and Harragan through the maze of corridors that linked the various offices that made up the nerve centre of the Thomas Cook Airlines operation. Eventually, we arrived in a room devoid of character, devoid of charm, the definition of functional.

As soon as we sat down, Burke announced she wanted to record the meeting. It was as if somebody, not present, wanted to know, what was said, by whom and when. Surely it wouldn't be the same individuals who had listened in to my conversation with Thorington? It wasn't difficult to understand why they would want to pore over the recording. I suspected it would form part of a concerted effort to twist and contort every word, every syllable and to take out of context anything I uttered. All this to get to, what I assumed would be their desired endgame; a disciplinary followed by a dismissal.

Those years in the music business had not gone to waste and had stood me in good stead to know bullshit when I smelt it. It had also taught me how to deal with it.

I told them I had no issue with them recording. I intended to record it myself.

Predictably the meeting started with the formalities. Burke confirmed this was a fact-finding meeting. She also confirmed that the allegation was that I had refused to fly the duty of the seventh of May. Interestingly, there was not a single mention of the illegally planned duty.

She was fixated. She was fixated on who said what to whom on the most distorted, unintelligible and least reliable communication mediums we use, the HF radio. Nothing else existed in her investigation bubble. Whenever I pulled her away to talk about the facts, hard facts, indisputable facts, her neck would turn ruby red again. Like a child separated from her security blanket, Burke would grasp out for the frayed edges of the HF call, only to relax when reunited with the topic I suspected she had been told not to veer from.

For the moment, this was Thomas Cook's train set. Burke could run this meeting any way she desired, and Thomas Cook could take any action they wanted, at least while this issue remained within the company's internal process.

We talked about the HF conversations to death. If she didn't like the answer, she would ask precisely the same question but in a slightly different way. She got the same answer every time.

Now maybe we can talk about the illegally-planned duty? But no, Burke was just not interested in going down that particular road.

Maybe this was not an aspect in which her masters had briefed Burke? I would have been surprised if she even knew where to find the requirements for a duty to be legally planned, never mind what those limits would be.

It brings me to a simple fact. In my experience, most cabin crew do not understand the legalities of flight duty periods, and few, if any, have any idea where to find the information. Most crew, if they were honest, would be the first to admit that statement as true. Again from experience, most have little idea about their responsibilities with regards to FDP and fatigue issues.

Burke was ex-cabin crew. It was, therefore, no surprise she wanted to go nowhere near discussing the FDP breach.

I did manage to drag Burke away from the damn HF transmission and steered the discussion onto FDP issues via subterfuge. I explained that I had experienced a similar situation three years previously, where a flight had been planned, in advance, in breach of the maximum flight duty period. Again the company had expected me to cover this known breach by utilising my discretion. I recounted that in this particular episode, once again, I was put through to Thorington, the same individual involved in the current case.

Burke feigned indifference. Suddenly she became particularly interested in her own notes, reading them as I talked. She did not want to know, and she did not want to hear what I had to say. I carried on regardless.

I explained that I had been rostered to operate a flight to Goa, returning a few days later. Her eyes glazed over, and there was that red tell making an appearance again. It was pointless going into details and expect Burke to engage. She wasn't interested, and she certainly wasn't listening. She was already working on how to ask the same question for the umpteenth time about the HF transmission. It could not have been any more transparent that this was all an attempt to elicit the answer she wanted. What she hadn't realised was, it was immaterial how she was reacting. The interview was being recorded. That's precisely why, irrespective of Burke's disinterest, I had introduced this information. I was already thinking ahead.

I had concluded, within the first few opening moments of the meeting, as far as Burke was concerned, trying to explain and reason with her was like banging your head against a very thick brick wall.

Crucially, Burke had failed to appreciate the significance of the disagreement between myself and Thorington concerning the Goa incident. The outcome of this disagreement directly influencing my stand in refusing to use my discretion the morning of the sixth of May.

If Burke had done what she was required to do and listened to what I had to say, she could have turned around to her masters and told them that there was no case to answer. There again, the fact that she wouldn't was precisely why I had assumed Hutchings had put her in that role in the first place.

What Burke had failed to appreciate was that Thorington had previously misinformed to me about the legalities of flight duty periods.

I had challenged him over the return leg from Goa. He told me it was legal. I told him I required it in writing before I departed. Thorington emailed stating:

"As discussed, the use of unplanned discretion due to unforeseen program disruption is acceptable by the authority for discreet crew rotations (inc. subsequent sectors) This although not common, is by no means uncommon."

Not common, but is by no means uncommon?.. Interesting.

Acceptable to the CAA as well, apparently, well at least according to Thorington.

The only issue there was, it wasn't. It was relayed back to me at the time that the CAA had subsequently informed Thorington that, by 'misinforming me' of the legal situation he had *'overstepped his authority by a significant degree'*.

Burke looked up from her notes and continued to bang on about the HF communication call. She had become white noise herself. I filtered her out. Anything and everything she now said, stripped of any relevance.

The red tell was now a constant blotch on her chest. She was getting nowhere, and the blotch told me she knew it. How would she explain this car crash of an investigative meeting to her superiors?

Suddenly she steadied herself. It was time to throw her ace on the table. 'We have actually asked for the recordings from Stockholm.'

'I would have thought you would have done that already', I replied.

Burke and Harragan looked utterly deflated.

Poor old Burke. Her lack of aviation knowledge was palpable, as was her understanding of what constitutes good airmanship. But there again, how would she know? She had never flown a commercial aircraft in her life. Given what I had seen of her performance so far, I suspected when she accepted the position, she had not considered, even for a moment, the role required a little more than being a uniform monitor and ordering the paper clips.

Did she not fully comprehend the optics to an impartial observer that, with virtually no relevant experience or background, she was heading an investigation into airmanship decisions by an experienced commander? Unless this investigation was conducted with absolute integrity and impartiality, it was a situation that could ultimately finish my flying career, finish hers, and cause yet another PR disaster for Thomas Cook. Again, given what I had witnessed during the meeting, I seriously doubted she had the imagination to consider the consequences.

Her lack of interest in facts only matched Burke's lack of aviation experience. She had her own catchphrase for any time I had attempted to explain anything. 'We will deal with that later'.

If she was attempting to stitch me up, she was doing an appalling, ham-fisted job of it. I was almost beginning to feel sorry for her. Almost.

Was it that unreasonable to expect a pilot manager, assuming they had any ability or integrity, to accept a pilot's word if he reported he was unfit to fly? It was beyond my comprehension how any company could trust an individual with a piece of equipment worth hundreds of millions of pounds, with hundreds of lives on board, yet not trust that same individual to know whether he was fit to fly. It appeared this was simply a staggering level of arrogance and / or incompetence.

Again I tried to drag her back to the reason for my decision.

Things were getting a little tricky for our Jo. The red blotch was now a constant reminder that things were not going her way.

'Ok. If you don't mind, I wouldn't mind adjourning. I just want to make sure that we have got everything that we want and require.'

*'Everything we want and require'*, well that gave the game away. It was nothing to do with flight safety. It was nothing to do with an investigation into the truth. This looked like it was nothing more than a setup. All she had caught on this fishing expedition was a pram and an old boot.

With that, we had a break. As Burke and Harragan left the room, Richard looked at me.

'Bloody hell Mike, you were right. It's exactly as you told me, there is nothing else. What the hell are they trying to do?'

Fifteen minutes later, we were back at the races.

Richard could see what was going on, and of course, he would report this back, not only to the CC but also BALPA HQ.

I opened the second half of the meeting by reading out a report from Thomas Cook's own aviation medical advisor Dr Hudson.

"Pilot fatigue from long duty periods, unpredictable work hours, circadian disruption and sleep restriction is now established as a major contributory factor in aircraft accidents." Later on in the report, Dr Hudson also stated:

"If accidents are only the visible tip of the iceberg, then pilot fatigue is a major cause for concern as a potent risk factor in accidents waiting to happen."

Nothing, apart from blank stares from Harragan and Burke. They did not have a clue how to respond.

What I was saying appeared to be going so far over Burke's head it didn't even cause her hair to part.

I also took this opportunity to detail the ramifications of the medical situation I had to deal with over the weekend relating to my children's mother. I mentioned that she was rushed into hospital with an undiagnosed and life-threatening illness. I explained I was not prepared to go into details of the specifics as it was no business apart from the family.

Did that go in one ear and out the other as well? Ultimately, I guess we will find out.

Considering Burke's evident lack of knowledge on the FDP issue, I explained the legal aspects relating to not flying when fatigued.

The effects of fatigue are closely linked to those of alcohol. It was vital that she, and therefore the company, were made aware of this link. Thanks to the recording, so would anyone else who might eventually read the meeting notes. I was making sure there was no hiding place for Burke to subsequently say "no one told me," or "I didn't know."

From my perspective, Burke appeared to be so beholden in attempting the stitch up she did not even feign interest in what I had said.

Ok time for a little fun, let's see if I can get that red patch flashing again.

'If I could ask a couple of questions?' It was rhetorical. I did not wait for an answer.

There we go, slight reddening.

'Does the company wish for pilots to report fit when they suspect they are unfit for duty?'

The only answer to that question should be a straightforward, 'absolutely not, and it would be ridiculous to suggest such a thing'.

Burke was caught like a rabbit in the headlights. Burke looked over to Harragan for help. The body language back from Harragan said 'you're on your own with this one'.

Burke, not for the first time, became visibly flustered.

'I'm not going to comment on that at this moment in time.'

I bet you're not.

I think it called for another one. I was enjoying this.

'Does the company wish for advance notification of possible unavailability to allow the maximum opportunity to re-crew?'

I knew the answer to this before I asked it, but, there again, so should she. The answer is in the airlines own documentation. Again, it all got a bit too much for Burke.

The correct answer was, as stated in the manual, you must give as much notice as you possibly can. If you don't, disciplinary action might be taken against you.

The red tell was now glowing like sunburn on a ginger holidaymaker. I just sat there and waited. She was drowning. Eventually, after much drinking of water and coughing, she ran out of things with which to prevaricate. I just sat there; arms folded my eyes boring into the back of her head.

'Again, I wouldn't want to answer that question.'

Unmoved, I continued.

'Do you think it is reasonable rostering practice to program two early morning sims followed by the duties that subsequently require discretion which was notified to the company ahead of time?'

Finally, the mask slipped.

'I am not qualified to answer that.'

What a fascinating statement to make from someone tasked with investigating a refusal to fly due to fatigue issues. It didn't tell me anything I did not already in order to come to a conclusion as to the level of Burke's incompetence.

My considered opinion was that Burke was just a pawn. A pawn that had not quite worked out she was being used. Was she a mere 'cherry on the tree', utterly disposable and utterly replaceable like an X-factor finalist? A wannabe manager instead of a wannabe famous.

If she thought this had been bad, it was about to get a whole lot worse.

I asked the question. 'What is the next stage?'

Burke glanced over at Harragan, the HR note-taker.

She explained that she didn't know what would happen next and she didn't know who would chair any possible disciplinary meeting.

Dark pink.

Richard joined in. He knew what I was getting at.

'It will be probably better if it is someone who is not directly involved.'

Again furtive eye contact between the two women.

Burke tried to contain what she was about to say, but the words forced themselves out, she spluttered:

'Well, they can't.'

Bright Red.

Harragan tried to help her flustered colleague.

'No, it will be somebody independent who has not been involved.'

*Come on, take the bait—Burke nibbled then she bit.*

'Anybody who's been involved that I've had any meetings with or anything like that will not be involved.'

Perfect. Mission accomplished.

Burke's chest now looked like someone had fired a high-velocity strawberry at her.

There it was, on record. Burke had said anyone she had met or who had been involved in the investigative meeting would not be able to hold the disciplinary meeting. Considering the only person who had been provided with all the details relating to my case had been Scadeng, Burke had just ensured that he would not be able to chair the disciplinary.

Oh dear, she was not going to be popular with her betters when they find out how badly she had just screwed up and stymied their intention.

I then suggested that, as this was a fact-finding mission, and Ian was a flight time specialist, he should now be invited in. She reluctantly agreed.

Ian wasted no time in saying pretty much everything Burke did not want to hear.

'So I think in terms of his commitments to the company and to himself he has to decide whether he is fatigued, whether he is potentially fatigued. The ANO and even your Ops manual states that's what he *must* do.'

Ian then spent about ten minutes dismantling her position before highlighting precisely the real issue throughout the industry.

'Can I just add something on the fatigue-reporting side? What is becoming very clear in the industry as a whole is that there is the widely-held belief that fatigue is being under-reported by pilots. Certainly, the CAA is saying now that they want to encourage fatigue reporting because it is the elephant in the room as it is severely under-reported.'

I suspected by now Harragan and Burke were having second thoughts on the wisdom of recording the meeting. They couldn't wait to conclude this encounter.

For the drive home, the sun was out, and life was great. I had just run a coach and horses through Thomas Cook's attempt to stitch me up. Any sensible organisation would, by now, realise they were stuffed. Surely they would understand, at some point, they were going to get a severe kicking around the car park, and if they did not back down, this was likely to be a very public kicking? Now, however, it would take more than a handshake and a pint to sort it out.

On two previous occasions during my flying career, I had

experienced situations that I had highlighted that, if not dealt with appropriately, would result in fatalities. The tragic and preventable deaths in Edinburgh and Charles De Gaulle bore testament that my concerns had been ignored. This time no amount of threats or mishandling of the situation by the airline was going to deter me from seeing this out. Without being too dramatic, quite literally, lives depended on it. It looked like I had the backing of BALPA, but even now, I was still giving the company the chance to concede they were wrong before I resorted to informing the aviation police, the CAA, about what was going on.

# CHAPTER SEVENTEEN

## *Professional Airline?*

*"The claimant confirmed that he was not, at any point, too tired on 6th May to operate discretion."*

Judge Slater

And then came silence, for over a month.

My time suspended was time well spent. The garden was looking stunning. The lawn was wearing stripes, even if my shoulders weren't. My one nod to a mid-life crisis, a racing kart, had laid in its constituent parts for years. Thanks to the enforced holiday, and help from my son, Jake, it was running as sweet as a nut. Unfortunately, the driver was more Lewis Carrol than Lewis Hamilton.

Finally, a month after the meeting, I was informed by Burke of the predictable outcome of her investigation. I would be subject to a disciplinary meeting that potentially would end my career as an airline pilot. An inquiry that I was told would take no more than two weeks had taken over two and a half months. Yet, even now, I was still none the wiser on precisely why Burke wanted to proceed.

There was no date and no documentation and therefore no idea when the disciplinary would take place or who would chair it.

I could only assume the delay was due to the various individuals involved finally realising they had a significant problem.

Their problem was this. On every single aspect, I had followed, to the letter, all legislation and company requirements. I was bombproof, with or without BALPA support.

That's the problem with stitch-ups. Facts can get in the way. I could think of no circumstances where any of my actions would warrant any form of disciplinary. Bizarrely I was now actively looking forward to seeing what depths Burke would descend to in her ill-advised attempt to discredit me and impress Hutchings.

I had thought that Thomas Cook would have come to their senses and realised that the longer this went on, the more likely it would result in a PR disaster.

Had Thomas Cook even considered the PR repercussions? Had they become so blinkered that they were lashing out without considering the dire consequences? Unless they did back down, the premise 'all press is good press' would be stress tested.

If things had gone into slow motion before, it had now become glacial. It had become a constipated process.

I put into effect an irreversible action. It was time to insert a regulatory laxative.

I had given the airline more than enough rope to climb back up and deal with the situation appropriately and internally. It was now time to let the company swing by it.

I informed the regulator.

My first port of call was a CAA flight ops inspector by the name of David Campbell. Although he was not Thomas Cooks' flight ops inspector, he was recommended as a trustworthy individual by a senior training captain who was appalled by the events.

The same training captain had serious concerns that if I approached Thomas Cook's own CAA flight ops inspector, there was a possibility my confidential report of the situation would somehow find it's way back to Hutchings. There was also a concern of potential 'regulatory capture'.

Proof that a regulatory merry-go-round did exist between regulator and airlines was evident in the recent appointment of a new director of flight safety at Thomas Cook. This individual had previously been an ex CAA flight ops inspector. Was this why it appeared Hutchings considered he could get away with just about anything he wanted? The very definition of regulatory capture.

I rang David Campbell. He came across as a very personable chap. I explained to him precisely what had happened.

He was understandably taken aback. He suggested that I write to him officially as a whistleblower at which point the CAA would investigate my allegations. I also explained, apart from being suspended for refusing the duty, I had no documentation relating to the investigation. Thomas Cook had refused to release the flight plans, and Scadeng had refused to answer my email.

We agreed that once I did receive any paperwork, I would forward it on to him. That should loosen things up nicely.

Another month passed, still nothing from Thomas Cook. Even now, desperate for crew and still paying thousands to sub external aircraft, here I was, a fully-functioning airline captain, with my feet up over three months later. My colleagues getting hammered, numerous tales of falling asleep during flights and near misses on drives home. My biggest issue was how to keep my beer cold in the garden. It was a beautiful summer. I was starting to believe that doing the right thing had some significant rewards.

So why would they be taking so long to deal with this issue?

I wondered if a court date had been set for the alleged taxiing incident yet. The two were not connected at all, were they?

Finally, on Aug 28th at 12:15, a knock at the door announced the arrival of a large package. Sixteen weeks after being suspended, I now, at last, had something to get my teeth into.

Up to this point, I still had no idea how Burke was going to square away, turning a captain complying with all legal requirements and Thomas Cook's own policies and procedure into a disciplinary. Finally, I was now going to find out.

The first thing to fall out of the package was a letter dated Aug 26th.

The letter informed me that in less than 24 hours, Thomas Cook planned a disciplinary meeting. At midday to be precise.

Despite the severe nature of the situation, I laughed when I saw the claims. I doubt very much that was the reaction Thomas Cook intended, but that was the reaction it got.

The letter setting out the accusations had to have been written by someone either clueless, desperate or maybe both. God forbid anyone who was a pilot had any involvement. The wanton ignorance of legislation, common sense and the lack of any apparent understanding of a commercial pilot's decision-making process was pitiful. This lack of knowledge would embarrass a student pilot yet to go solo.

BALPA HQ and the CAA were going to have a field day taking this apart.

The disciplinary letter posed several questions. Not least of which was:

If Burke had completed the investigation, why had Harragan, the HR note-taker, signed, what in effect was, Burke's conclusions? There were only two reasons I could think of. The first: late in the game,

Burke realised she was being stitched up, and, sensing where this matter was heading she belatedly wanted to distance herself from what was to follow. The second was, I suspected Thomas Cook had used her as the fall guy. Now they had got the *'dodgy dossier'* out of her she was of no further use.

My gut feeling it was the latter.

Along with the letter was a blue ring binder crammed to bursting with the investigation report, interviews, emails, and other assorted documents all with one apparent aim. To destroy a reputation and career that had taken two and a half decades to build.

Sixteen interviews required for a who-said-what-to –whom-and-when on an HR call. All because I refused to commit a criminal act by agreeing to fly when I suspected I would be fatigued.

As I read the letter and then the contents of the folder, I became a little offended. Not because the letter stated I was dishonest and lacked integrity, which it did, but because Thomas Cook obviously considered me incapable of tearing down their straw-man arguments. Maybe it was a case of them overplaying their hand? Perhaps they believed I would not fight the case considering my age and seniority would restrict my ability to find a similar position with another airline. As a tactic that had worked previously with the pilot workforce, had they made a mistake, thinking it would work on me? Maybe they just believed the old boy network that Scadeng had left behind at the CC and BALPA would serve them well? The way the CC, apart from Richard, had cast me adrift, tended to support my suspicion.

Perhaps Hutchings considered the CAA would not rally to my cause given that his new director of flight safety buddy had friends in high places given his previous role at the CAA.

The *'Invitation to the disciplinary hearing'* letter stated who was going to chair the meeting.

It was none other than our friend Scadeng himself.

Now wasn't this interesting?

Had Burke interviewed Scadeng during the investigation? If she had, according to Burke, he could not chair the disciplinary. If he had *not* been interviewed, then any claim that this had been a 'thorough investigation' was shot. Scadeng was the one person who I had sent all the details of the matter to months earlier. He was 'involved' because I involved him. This removed any ability for Scadeng to claim 'plausible deniability' on the issue. It was almost like I knew what was going to happen before it happened. Maybe I did.

Chairman Scadeng wanted me to attend a meeting less than twenty-four hours after I had received hard copies of what I was being accused of.

Not a chance. I told him why in an email;

"Today at 12:16, I have received through the post the documents relating to a case from May 7th. A swift perusal would indicate that the company wish for a meeting tomorrow at 12:00L, i.e. in less than twenty-four hours. It would be unprofessional for me to commit to that meeting without having time to peruse the documents and discuss the contents with my legal advisor and BALPA. I would have hoped as a supposedly professional airline the company would have given more notice than it appears to have."

I also drew his attention to how long this issue had taken compared to their stated target adding:

"Why, if all the details and interviews were completed by the end of May and mine June 12th, has it taken the company nearly four months to provide me with details of what I have been accused of, and the basis of that accusation? Why have I been given less than twenty-four hours to prepare?"

The last paragraph was a shot across Thomas Cook's bows that there would be ramifications if the case proceeded. The whole tone of my email was to make it abundantly clear I had no intention of backing down.

It was not a threat to the company. It was a promise of what was to come.

*'Please also be aware that this email will form part of any 'bundle of documents' should I be required to pursue this matter through the court process.'*

My next move was easy. I would go and sit in the garden and eviscerate Burke's attempt at a hatchet job.

It was a good job I was suspended. I was going to be very busy over the next few weeks.

Goodness that sun was hot.

# CHAPTER EIGHTEEN

## *The Gental Way*

*"The claimant argued consistently that he considered he was not legally entitled to exercise discretion where, in his view, the flight had been illegally planned to exceed the maximum FDP."*

*Judge Slater*

In life, it is an infrequent occurrence to know you are one hundred per cent right on anything. No doubt, no uncertainty, just completely confident you are correct. This was one of those rare occasions. It was one to be savoured, maybe even one to enjoy. It was not bravado or an attempt to be cocky. It was a quiet, unwavering belief that I could do more damage to them than they could do to me. The more they pushed, the more it was ultimately going to hurt them.

Reading the notes from the interviews and investigation, the more Thomas Cook managers, crewing and operations staff embellished their stories, the more they became mired in their own deceit. It was like watching a fly captured in the web, the more it fights, the more the web wraps its sticky restraints tighter. I was the web.

The sun was out yet again, too hot to sit in for long. It was time to get the sun parasol out. The cast iron table in the middle of the garden was to be my office for the next few hours. My fully-charged laptop sat in front of me. The cursor blinked expectantly at the head of the first of the PDFs that collectively made up the electronic version of the investigation—winking as if inviting me to start attacking Burke's fiction with a flurry of red text. To the left of the laptop was the file, the hard copy, to the right a notepad to jot down items and ideas to refer to later. The refreshment of choice was the family's first attempt at cider-making. In its raw form, it had similar characteristics to the blood from

'Alien'. Topped up with blackcurrant, however, it was fit for human consumption, just, and it did prevent the roof of the mouth being dissolved. It was described by my son, Jake, as having a certain 'zest'... Ok, to work.

Before I hit the investigation folder, I read for the umpteenth time the disciplinary letter.

Harragan, not Burke, signed the letter containing the allegations. That in itself was deeply suspicious. Why would a note-taker and not the manager sign a letter stating the findings of the inquiry?

Without any information to the contrary, it would be reasonable to assume the letter was Burke's pronouncement on the results of her investigation. It would, of course, be out of the question for anybody else to be involved. This matter was confidential, after all. I reasonably concluded, there was only one possible explanation. If Thomas Cook had abided by their own policies and procedures, this was Burke's work, then Harragan signed it.

When I had attended the investigation in June, Burke had pronounced:

'We're here to investigate an allegation of refusal to operate a rostered duty which was the TCX2684/2685 on 7th May 2014.'

That was it—*nothing* more and nothing less.

Mysteriously though, another accusation had, not-so-subtly, been slipped in the disciplinary letter. It was one that had not been raised at any point previously:

"In protest at the flight having, in your view, been rostered to exceed the maximum FDP, you told operations that you refused to work 15 extra minutes' discretion to land the aircraft at its destination."

The letter concluded that the refusal to operate the fifteen minutes of discretion and failure to perform the duty on the 7th were both gross misconduct due to '*refusing a reasonable management instruction*'.

At the foot of the letter, it stated:

"If you are found guilty of gross misconduct, you may be dismissed without notice or pay in lieu of notice."

I could only assume that, as Burke had headed the investigation, it must have been herself who had come up with these eccentric conclusions.

They were eccentric because they made no sense in any capacity whatsoever.

There were only two conclusions I could come to. Burke was either utterly clueless as to the legislation, rules and regulations that constrain what a commander can or cannot do, or, she was complicit in a disgraceful attempt at an underhand internal investigation.

What was not in doubt, Burke was either arrogant or delusional enough to challenge a 13,000-hour, highly-experienced airline captain on the intricacies of aviation law relating to maximum flight duty periods and captain's discretion. Alternatively, she was desperate to stay on the 'right side' of someone already described to me as a 'nasty bastard'. Maybe her own position depended on it?

If that was not ludicrous enough, Burke then, with no medical training, at least that I was aware of, pronounced that:

" On 7th May in protest at your flight on 6 May 2014 having, in your view, been rostered to exceed the maximum FDP, you dishonestly stated that the reason for refusing that duty was fatigue."

This whole sorry situation was of Hutchings' own making. He owned this calamity. He decided to put individuals with the mentality of Burke in pilot management roles. From the tone of the letter and investigation file, it was not difficult to deduce why Hutchings had made such a decision. From the evidence provided, I reasonably concluded that Burke's primary role was to adhere to whatever she was told to do, think or say. No questions asked; no room for independent thought. The accusations in the letter were the inevitable outcome of this dynamic. This was the culture that had been imposed on Thomas Cook since Hutchings had taken charge. How anyone of even average intelligence could conclude that the disciplinary letter was the result of a reasoned and unbiased thought process was, to my mind, quite frankly, pathetic.

There were so many issues it was challenging to know where to even start. But, start I did.

The first accusation was, "you refused to work 15 extra minutes discretion". The letter stated that this amounted to "refusing a reasonable management instruction".

There was only one way to interpret that statement. Thomas Cook airline management believed they had the right to '*instruct*' me to use discretion. Had they been sniffing the felt tip pens ?

Burke appeared utterly oblivious to the fact that it's referred to as 'Captain's Discretion' for one reason. It was a none-too-subtle hint as to whose discretion it was. I was not sure how Burke could become so confused that she was under the impression that she, or any other Thomas Cook manager,—never mind an ops controller—could have any say, or valid view or input in its use or indeed its refusal. They might yearn that they did, but they don't. It's the commanders. It was mine and mine alone.

This was akin to facing a disciplinary because Hutchings had asked me to lend him fifty pounds, and I had refused. It was my money to give; it was my discretion to grant. It was in my gift to say yes when legal and my right to say no at any time for any reason. I did not even need to give a reason.

The other small problem Burke had was this. I did use fifteen minutes of discretion. When I legally could.

It was not the greatest start to a disciplinary letter.

For Burke to claim it was gross misconduct requiring a disciplinary, explained everything that you needed to know about Ms Burke and the Thomas Cook management mindset. In my opinion, one word summed it up:

Delusional.

It got better. The letter then claimed that, due to 'refusing' to operate into discretion, this would "force the aircraft to divert, resulting in the potential for increased risk to safety".

Was Burke's overactive imagination triggered by watching one too many disaster films? Was she seriously claiming that a diversion would potentially increase the safety risk? Why would she think landing in Gatwick was more hazardous compared to Manchester? Did she know something pilots don't? Was Gatwick airport mined? Did a youth centre now run ATC? Did she portend the drone issues?

You could almost see the desperation dripping off the page. Actually, it was my perspiration... damn it was hot—time for another cider.

It was the second accusation that was the undoubted winner of the 'what planet is she on?' stakes. It seriously questioned what Burke had spent the last sixteen weeks doing. One thing was sure; she had not spent it studying the legislation and Thomas Cook policies and procedures concerning fatigue. If she had, she could not have written;

"7th May in protest at your flight on 6 May 2014 having, in your view, been rostered to exceed the maximum FDP. You dishonestly stated that the reason for refusing that duty was fatigue."

Was Burke a medic? Was she a sleep expert? Had she even consulted with anyone with a medical background? How exactly would Burke or indeed anyone else know if I was or was not fatigued? The only person who would know would be me.

Any lingering hope that this was all a mistake had now evaporated.

Thomas Cook had now gone through my V1 point of giving a crap about what Burke's or the company's view was or what they thought about me. They collectively appeared to be one colour short of a rainbow.

It was time to turn my attention to the full lever arch file containing the interviews, numerous internal email threads, along with the investigation report itself. I concluded the investigation document was either the product of a confused, dishonest, or easily-led mind. I honestly could not think of any other possible explanation. It also appeared the company was attempting to drown me in paperwork, trying to muddy the waters.

A straightforward rule I learned from my days in the music business was that when someone is lying, they overcompensate. They know they are lying. They know that you know that they are lying. The Pinocchio effect manifests itself by embroidering every issue, expanding every point, attempting to cover the dishonesty with intricate detail. Did this explain the reason that Burke conducted sixteen, yes, SIXTEEN interviews? All this effort for three phone conversations and one HF call I made on May 6th. If this was not overcompensating, then, personally, I don't know what is.

Information passed to me by colleagues who had previously endured disciplinaries with Thomas Cook confirmed my suspicion. Thomas Cook leveraged the fact that few, if any, pilots or cabin crew had the stomach to fight any disciplinary action levelled against them, no matter how unjust. The crew members involved knew, with good reason, it would cause serious complications for their future career and future relationship with Thomas Cook. Why would the airline bother spending time building a solid case when employees would usually just fold anyway? As far as Thomas Cook was concerned grievances and disciplinary issues were a fait accompli before they even started.

That was until now. I would not be backing down.

The fingerprints of senior management were all over this. They appeared to have become emboldened by the toothless response from BALPA over numerous contentious issues and the lack of resolve by my fellow pilots and cabin crew colleagues to fight any case thrown at them.

Hutchings and his poison posse, however, looked like they had made the same crucial mistake that senior management make in many different industries. They had started to believe that they were infallible. They had begun to believe their own hype. The simple fact is, if you surrounded yourself with sycophantic underlings, the offices then become an echo chamber. Management is told only what they want to hear. The weak-willed, too afraid to challenge directives or decisions, are then promoted, only serving to further exacerbate the problem. This further entrenches the manager's view that because they say something is right, then it is right.

This is referred to as confirmation bias. The definition is "tendency to search for, interpret, favour, and recall information in a way that confirms or strengthens one's prior personal beliefs or hypotheses".

As an example, what, other than a vastly inflated, self-delusional mindset, would allow any manager to feel confident enough to claim that a third person was dishonest when claiming they were fatigued? This was the equivalent of saying somebody was lying about having a headache. In my opinion, Burke was never interested in the truth. She was only looking for information that could be twisted to confirm, to what I suspected was her preordained, pre-investigation, conclusions. I strongly suspected she was told by her superiors what the outcome would be before the first interview ever took place.

I paused to consider the following. Ultimately it would lay the foundations for the tone of my response.

I reflected on the journey I had taken since my conversation all those years ago with Andrew Gold in 1986 over a beer. The collapse of the record label distributor, leaving me penniless. The day sheltering from the rain in Waterstones in Manchester, and picking up the pilot study book. The weeks of travelling on the bus to Manchester Central Library to exchange the last 'Ground Studies For Pilots' for the next in the series, having no idea how I would fund this journey to become an airline pilot. The days and nights spent poring over study books and the cost of training, leaving me tens of thousands of pounds in debt with bank loans and credit cards maxed out. The years earning a pittance as an instructor and having to travel all over the world to fly. Finally joining an airline and eventually becoming a captain on wide-bodied jets flying across the globe. The millions of miles I had travelled and the hundreds of thousands of passengers I had flown safely from one country to another. All without ever putting a foot wrong. All without placing any passengers in danger and all with zero challenges to my professionalism and integrity. An exemplary sickness and disciplinary record to boot.

And yet.

And yet, here we were. Burke, a mouthpiece for her superiors, imagined she could ruin a pilot's career and reputation; a reputation I had meticulously built since 1989, without any repercussions.

Pilot manager? I wouldn't trust her to make the tea.

Burke and Thomas Cook had no idea what was coming over the horizon. This was going to get messy. The gloves were coming off, and the Doc Martens were being laced up.

I considered my options, my tactics. I had one absolute. I would not accept those accusations or any action taken by Thomas Cook as a result of those allegations.

So, what was I up against? It looked like the entire management structure of a multibillion-dollar, internationally-recognised conglomerate. A management belief that they could steamroller over anyone and anything. It was clear they were not interested in facts. It was clear they had no intention to be reasonable. I would have to think smart; it appeared *they* were incapable of it.

Before I realised, I looked down at my note pad. I had idly scribbled 'judo'.

In my youth, I had practised judo. It had taught me a lot and not just about discipline.

Judo was founded in Japan in 1882 by Kanō Jigorō. It is a martial art and also a way of life, a philosophy. In Japanese, it means *'gentle way'*, and is built on the principle of using the strength of an opponent against them. If they push, instead of pushing back, pull, and help them on the way to the floor.

My mind went back to when I was around fourteen years of age. I had been practising the 'gentle way' for approximately six months.

It was announced during an assembly that one of the teachers would be holding a lunchtime 'judo club' and all were welcome.

The very first meeting was a week later. Ten of us were sat in a semi-circle around 'Sensei'.

The teacher explained that judo was a valid form of self-defence, and there were techniques on how to combat attacks.

'Mike, come up here and attempt to throw me.'

Here was I, a skinny, ginger-haired kid with biceps like knots on cotton, up against a fifteen stone 6ft 2' teacher. He wanted me to throw him, did he say?

I had already decided what to do.

I immediately went in for an O Goshi. This is a 'hip throw' where you slide in front of the individual you are trying to throw, grabbing their jacket, turning your back to them and heaving them over your hip. The size and weight differential made this almost impossible to achieve. It was misdirection. I waited for the exact moment I could feel the teacher pull back to resist the throw. At that precise second, I spun around whipped my right leg behind his and helped him on the way to the mat. He went flying through the air. It was the most perfect throw I had ever, and *would* ever achieve. It also was the easiest. I just used his own inertia against himself.

My plan now was simple. My reply would be quiet, controlled and structured.

My intention? To help Thomas Cook managers destroy their case by using their own perceived strength. They would see my response as trying to pull them one way. The reality was, my plan was to go another.

The first step was to produce a bundle of documents to present to Scadeng in advance of the disciplinary. The key would be the covering letter. There would, of course, be a full rebuttal of the accusations. My only concern at this stage was that the company would then see sense and call a halt to the madness. It was a judgement call, and I did not want to trigger reciprocity. I was not interested in 'doing a deal' by this stage. I wanted to see them, like the judo teacher, on the flat of their back, embarrassed and ridiculed. I had had enough. More importantly, it was in the public interest for this to happen. This was so any other airline manager who attempted to play the same game would see that ultimately they could not hide behind the anonymity of the company or claim they were just 'doing their job'. They would also see it came at a cost, a personal cost.

I had to trust my instinct. My instinct was telling me that the collective arrogance of all those managers involved would not allow them to back down. This would be as a result of the echo chamber currently installed at Runger Lane, Manchester Airport.

My in my, and many others opinion, the boil that was collectively Hutchings, Scadeng and Debus, needed lancing. The puss required drawing, and I had the needle. Debus was the CEO of Thomas Cook Airlines. Ultimately it was he who had installed Hutchings as DFO.

I was also up for a fight more than I had ever been in my life. I wanted to make this such a painful experience to all of those involved that it would make them think twice about ever doing this again to any pilot or member of cabin crew or office staff who was intent on putting safety before corporate greed.

I decided that an even larger lever arch file than the one sent to me by Thomas Cook would send the right message.

First I would demolish every accusation made, line by line. Then, within its contents, I would place, and prime metaphorical booby traps ready to detonate at the appropriate juncture and expose, what I considered, the devious and dangerous mentality of the individuals involved.

It would mean at least the doubling of paperwork. If it resulted in a hernia due to the weight of the documentation, so be it. It was a risk I was prepared to take.

I answered and then challenged, in red, every point that Burke had claimed. For added spice, I also included in the return bundle, an investigation into the pilots' lifestyle by a well-respected expert in the field of fatigue, Dr Simon Bennett.

It was at this point Lisa came into the garden to tell me she was going out and asked what did I fancy for that evening's planned BBQ, and was there anything else I needed. I explained I was going to need a large ring folder to hold this tsunami of paperwork.

First of all, I had to decide what colour the folder needed to be. I needed to take several aspects and features of this disciplinary into account before finally settling on this critical issue.

Scadeng had decided he was going to throw his weight around in chairing the disciplinary. I was in no doubt, in his mind, he was the alpha male. It appeared Scadeng was confident he could show Hutchings he was the silverback able to deal with *'difficult'* pilots.

So how should I present this bundle of paperwork? What does every silverback male chief pilot desire in a document folder? Would he prefer a reserved dark blue, or maybe a business-like black? I made the decision there and then. It would be a bright pink one.

Lisa did me proud. On her return, she presented me with a folder so outrageously DayGlo pink that it made the Eurovision song contest and Rue Paul's Drag Race look like an episode of Ultimate Cage Fighter. Yes, I am sure Scadeng would appreciate the thought, and I did not want to disappoint our new chief pilot.

Was this childish on my part? Of course it was! This whole issue was far too important to be taken seriously.

The red text had already started to appear in the investigation document.

I rammed home the unarguable fact that I had raised a simple question to Scadeng directly in my email two days before I was suspended. I highlighted that despite the passage of time, he had point-blank refused to answer any of the questions. Was this because he couldn't without blowing Thomas Cook's whole argument?

Despite the immense amount of paperwork, Burke did not see fit to include the most important evidence of all within the investigation bundle. The flight plans were conspicuous by their absence. The plans contained all the details of the duties. The planned sector times, the winds on the day and the turnaround times.

Due to Thomas Cook concealing this vital information, I had to work on my memory of the day rather than the documents that spelt out the exact details. Not ideal, but given the company's refusal to furnish these documents, I had no other choice.

There was another curious and interesting omission. Despite Burke stating in my meeting that they had asked Stockholm for the recordings of the conversation on HF, transcripts of the discussions were not included.

That was not all. Thomas Cook claimed every single phone conversation made to the company by pilots *'will be'* recorded. They did not say *'might'*, or *'maybe'*, but 'will' be recorded. How did I know that? Because every call to Thomas Cook would start with a recorded message reminding you of that fact:

*'Thank you for contacting the Thomas Cook operations and crewing information system. Please be aware all calls will be recorded for training and quality purposes.'*

Who said what to whom and when could, therefore, quickly be confirmed.

As a result, the red ink of my investigation into *the* investigation kept flowing. Or rather, the red text of the font on the laptop was becoming extensive.

Challenges to what it was claimed I said or what someone overhearing claimed to have thought someone else thought I said were aplenty. Consequently, my comments were liberally peppered with my new catchphrase *'Just listen to the tapes'*.

I was in no doubt. At the rostering stage, someone had to have knowingly massaged (def. from the Collins Dictionary to falsify, fiddle, misrepresent, fix, alter, cook) the figures to give the impression that the FDP would fit within the maximum time. If I could prove it, or someone admitted to it, that would be a criminal breach of EU and UK legislation.

It was never possible for these fictitious times to be achieved. Once the flight plans were produced, it would uncover the falsified times in the roster. That was why Thomas Cook would *never* release those plans. It would destroy their position.

I was in no doubt, Thomas Cook had assumed on the day in question that the I would not pick up on the illegality, or if I did, would have flown it anyway, not wanting to butt heads with management.

Again, I'll use the HGV analogy. It will help clarify.

Imagine an HGV driver has two hours left on their tachograph for that day, and the company is aware of this. The driver arrives at the haulage yard in Manchester. On the operations board, by the side of their name, is their duty for that day. The company wants them to drive a truck from Manchester to Glasgow. Having driven that trip many times, the driver knows it will take precisely four hours if they operate at exactly the speed limit, and there are no delays due to traffic. Despite this, the haulage company show the truck departing at midday and arriving at 14:00. They claim it will take two hours, to the minute, to drive. The company knows the driver only has two hours available on his tacho. When challenging that falsification, as it would be a physical impossibility to make that journey in two hours, the driver is threatened with dismissal, for refusing a '*reasonable management instruction*'. The company claim he still has two hours available on his tacho; therefore, he can operate.

If you now replace 'HGV driver' with 'airline pilot', 'tachograph' with 'FDP', you will have a much clearer idea just how ridiculous Thomas Cook's position was.

So what did Burke give as the reason the FDP breached the maximum allowed? Apparently, it was:

*"Due to a last minute change to the runway in AYT, it meant discretion would be used by 9 minutes."*

In Burke's world, a delay due to a runway change in Antalya in the afternoon was the reason, before I had even got to work that morning, for the breach in the FDP. Discounting a space-time disturbance caused by a passing black hole, it was beyond my grasp of physics, how Burkes claim made any sense, in any universe.

In reality, when the depths of the falsification of the flight times and the turnaround times were discovered, the violation was fifty-seven minutes.

My response in red said it all:

"FALSE. No it did not, the flight would have exceeded the maximum legal FDP due to it being planned, even at the maximum cruise speed possible of the aircraft, to exceed the legal maximum FDP."

In the investigation, Burke admitted that Thorington had informed crewing not to take me off the flight despite my telling him I would not be available. No wonder the airline was hitting the national press with cancelled flights.

Unwittingly, Burke then proved she was either devoid of understanding of the legalities associated with flight duty periods or, was prepared to claim evidence existed that didn't. She had falsely maintained that when I highlighted the illegal FDP, the company had produced a new set of flight plans that had reduced the FDP to below the legal maximum. How would she know that? Did she have had access to these high-speed flight plans? If so, where were they? Surely to goodness, she didn't merely take somebody's word for it without demanding to see the evidence. Given the implications, if these plans did not exist, no one would be that devoid of common sense and self-preservation, would they?

It was now becoming comical. Burke claimed that once we landed, Thomas Cook would have arranged a truck to tow the aircraft from the runway onto stand claiming the FDP would stop the moment we landed.

I kindly, in red, explained the basics to her:

"It would also have caused a major logistical problem at one of the UK's busiest airports. Furthermore, the FDP is from report time until the aircraft finishes 'on chocks' at the completion of the final sector. Therefore being 'towed' onto stand would achieve nothing."

Run an airline! Would you trust them to run a bath?

So how did Burke plan to go about her task to investigate someone who refused to break the law and put passengers' and crews' lives in danger?

She claimed to *'ensure I had all of the facts'* she interviewed all crewing and ops staff on duty and all management pilots who were mentioned in any interviews.

There had already been two significant omissions from the bundle, the transcripts of the HF and phone conversations and the flight plans. There was a third. It gave Thomas Cook's, and Burke's, game away as much as her red blotch.

Not only was Scadeng mentioned in interviews, including my own, he had also been sent the email. The one individual who knew most about the details of the case was Scadeng. By submitting the email on the morning of the 7th of May, I had removed Scadeng's ability to claim 'plausible deniability'. He was involved because I involved him.

Despite this, there were no minutes of any meeting or interview with Scadeng. Had he even been interviewed? If he had, did Burke remove the minutes from the file? Maybe, as evidence Burke was being used by Hutchings and Scadeng, she had included it, but Harragan, with the connivance of others, had removed the document without telling Burke? Whatever the reason, the intention was clear. It was to mislead and deceive those not able to chew gum and walk at the same time. It was an attempt by the company to claim Scadeng was not involved, and therefore could chair the disciplinary meeting. Not only can I walk when chewing gum, but I can also play triplets on a drum kit while masticating a juicy fruit. In other words, that bullshit did not wash.

I reminded them, in red ink, just in case they had forgotten, about every time he had been mentioned in emails and interviews with others. It was extensive.

I concluded "the chief pilot is involved in the situation. Therefore, he should not be in a position to chair the disciplinary meeting, as he discussed the matter with several other managers throughout the day."

That should put the cat amongst the pigeons.

This whole situation came down to the differences between myself and Thomas Cook regarding considerations of passenger safety and commercial concerns. My view was that a duty that expected me to land a Boeing 767, at night, into one of the busiest airports in the UK, after being awake for nineteen hours, in itself, would have put Thomas Cook passengers at serious risk. I would also have committed a criminal act if I had agreed to operate it.

Myself, and probably the vast majority of pilots' views, was perfectly summarised by Dr Simon Bennet:

In The Pilot Lifestyle: a sociological study of the commercial pilot, he stated that there was a disconnect between what an airline professes to be its safety commitment and adherence to just culture and speaking out policies that carry the CEO's signature and what actually transpires. In short, there is a disconnect between rhetoric and actuality.

In straightforward terms, the airlines expect the public to buy the line that a business, whose only real duty is to its shareholders, places crew welfare and passenger safety ahead of profit.

Burke had also included in the bundle, a printout that traced my predicted level of 'effectiveness'. It was provided by the FRMS manager Jane Marsden.

FRMS stands for Fatigue Risk Management System. It is used to monitor pilot fatigue using a computer program called FAST or Fatigue Avoidance Scheduling Tool. This system is used throughout the world, not just in the UK. Its purpose is to ensure that pilots, train drivers or indeed, any group of individuals whose abilities and fitness and alertness will directly affect safety, can be monitored.

The software has a straightforward task. It alerts the airline if its employees might be expected to be approaching such a level of tiredness that it amounted to fatigue. The data input for any particular individual will include the time of falling asleep, the time of waking, and how long the working day is. The FAST software currently predicts the performance of an average person on tasks especially sensitive to fatigue with the tool-making reasonable predictions. Along the Y-axis of the graph is the 'effectiveness' of the individual. The X-axis represents the time of day. As you would expect, the longer an individual is awake, the 'effectiveness' of that individual reduces. A reduced amount of sleep will also affect the level of 'effectiveness'. The system has its weakness, namely FAST only forecasts the effects of sleep and circadian rhythms on performance. It cannot account for other factors that alter performance such as environmental conditions like stress, or a variety of other variables which are known to be fatigue-inducing. Therefore, in effect, the FAST printout will always tend to indicate a higher 'effectiveness' level in an individual than is the case. In simple terms, the level of fatigue is likely to be worse than that predicted.

In my big pink file bundle, I was preparing to launch back at Scadeng, I again referred to Dr Simon Bennett. He stated that pilots would stop filing fatigue reports if they believe management will disregard those reports, or if they fear they will be victimised. His view was also that there was evidence of pilots reporting sick when they were, in fact, fatigued. Would this explain the reason that during my suspension, when the airline had an *'unforgivable lack of crew'* , the number of reported incidents of fatigue dropped, yet in the same period, sickness rates doubled?

I pulled no punches as I continued the red wave of text. The Doc Marten-clad foot was now being fully administered to Thomas Cook's nether region. I highlighted Burke had ignored the conclusions of Thomas Cook's own medical advisor, and had not made one mention of the ramifications of the dire family medical situation I had also been dealing with at the time. I suggested:

"Please draw your own conclusion as to why this important information was completely absent from the investigation report."

Fatigue is the killer, Thomas Cook's own manuals determined it's a killer, history proved it was a killer. So did the company's own Doc Hudson, stating very clearly that it was only a matter of time before fatigue results in yet another catastrophic incident. Did this hold any weight in Burke's deliberations? Apparently not.

Finally, the investigation report concluded:

"The decision-making process was not what would be expected from a commander."

That statement does beg the question, how would an individual who had never operated as a commander know what the decision-making process of a commander was?

Given all of the interviews, emails, evidence, and the investigation report itself, it was hard to come to any other conclusion what Burke expected. It was for Thomas Cook pilots to fly a duty, even if it was illegally-planned and agree to fly an aircraft even if it would be likely to endanger your passengers' safety due to fatigue.

Maybe Burke, with her lack of knowledge, could be forgiven for her ignorance in these matters. Perhaps she just expected that everyone would have the same herd mentality as she appeared to have and, as a result, would just *do what they were told.*

The investigation did, however, reveal that someone who should have known what the regulations demanded was the very individual who I had talked to when I had landed. Thorington stated in his email, included in the bundle that my "decision-making was not aligned to what our expectations of what the decision-making process would be from a commander".

Maybe that's where Burke had got her snappy one-liner from?

Considering Burke had also referred to 'my belief' that the flight plans were illegal, would it not make sense to present those flight plans to disprove that 'belief'? That is, of course, unless it would show that they were indeed illegal and it wasn't just a 'belief'. I knew they were illegal. I could think of no other reason for their absence.

It looked like a major airline, with access to legal experts, barristers, HR specialists, doctors, and the latest fatigue prediction software was intent on going head to head with one individual who could not even secure the support of his own union CC.

Richard had reliably informed me that for the last few months, Scadeng had been briefing against me. He mentioned the chief pilots old chums still held sway on the BALPA CC. Was it, therefore, any surprise I had received the cold shoulder from the union? Apart from Richard, I received not one call from any representative of BALPA. Despite this, I was still completely confident that ultimately I would prevail. I had the cavalry on my side, well the CAA—well to be exact David Campbell, the CAA flight ops inspector.

How could the CAA not support my position when their own publication states:

"It is in the public interest that those who comply with aviation rules have confidence that those rules are enforced.

Employees are entitled to look to the CAA to ensure that employers comply with those aspects of aviation related health and safety protection for which the CAA has been made responsible.

A failure to comply with any of the requirements of the Air Navigation Order and its related regulations or certain European aviation safety regulations is a criminal offence."

By the end of my review of the investigation bundle, the amount of red text had tripled the size of the original file that Thomas Cook had presented to me. I had covered every single rabbit hole that they could squirm down.

I had applied a filter to their attempt to muddy the water.

The last thing I needed to do was write the covering letter to accompany this significantly-revised bundle.

The letter to Scadeng would leave him and the company in no doubt what the ramifications of their actions would be.

I informed him that the bundle "may be required in the event that this situation, or any other action resulting from Thomas Cook's handling of this matter, is taken outside the environs of the internal Thomas Cook procedures".

I also let him know what was waiting for him if he attempted to screw me over. I advised Scadeng that there would be *'serious ramifications'* if he did.

I suspected there would be an inevitable, underhand and dishonest attempt to put words into my mouth, or intentionally, take anything I said out of context. I, therefore, suggested Scadeng very carefully read and digested the entire contents of the bright pink folder. I stated this was to:

"ensure that there is no misunderstanding, or mishearing, or disagreement during the meeting on what has been alleged, and my response to those allegations. My comments (in red), views, and interpretation of the ANO and other relevant documentation are all contained within these documents. Again, this is to ensure that nothing is 'missed' with regards to answering any questions or clarification of events presented in the documents that were sent to me."

I knew that this refusal to give an inch, and in anyway supplicate myself to Scadeng, would goad him into overreaching himself. I wasn't interested in coming to any accommodation with either him or anyone else for that matter. I was leading him to exactly where I wanted him. He just hadn't realised it yet.

The hunted had now become the hunter.

# CHAPTER NINETEEN

## *Hello Peter*

*"Fatigue has, for many years, been a worrying issue for pilots, and it continues to be a huge concern across the industry today. Pilots have told BALPA they believe fatigue is now the biggest single threat to flight safety."*

BALPA Website Home Page.

I had given Thomas Cook time to pull back from the brink. Collectively, it appeared, the management had elected to close their eyes and step over the edge. Maybe it was the last vestiges of a misplaced sense of loyalty that, up to this point, had prevented me from giving them a push. However, now I had the bundle; albeit, without the flight plans, it was now time to let the CAA know precisely what was going on. It was now time to give the suicidal base jumpers a shove over the precipice and put a hole in their parachute.

Due to concerns that the information might find it's way into the hands of the Thomas Cook management if I sent them to their own CAA flight ops inspector, my documents were only sent to David Campbell.

I explained to David the specific legislation that Thomas Cook had breached and that there was no room for interpretation.

I proposed that the only way left for the airline to see the error of their ways "would be in the full spotlight of the media and with the authority's involvement".

I suggested the reason for Thomas Cook's actions were due to the airline being "critically short of crew" and that their attack on me was designed to "stamp out any crew reporting fatigue".

I highlighted that the "sickness rate for pilots has shot through the roof. In other words, the real problem is being hidden".

I then included, in detail, a list of all my previous duties so that Campbell would be under no illusion of my claim to the extent of Thomas Cook's abuse of legislation designed to protect passengers and crew.

I attacked each of Thomas Cook's accusations and demolished every single one. The CAA would have no option but to drag the airline and individuals over the coals. I concluded:

"As you will have gathered, I am 100% confident that my actions were 100% correct, to the extent I am putting my reputation and livelihood on the line on this matter.

To summarise the situation, the airline's senior management have attempted to coerce an airline commander to knowingly commit a criminal act, and because I refused, I have been suspended for four months and have lost a considerable amount of money."

Four days later, I had my reply from David Campbell.

He asked was I happy for the information to be used to investigate Thomas Cook.

You bet I was!

There was, however, one potential issue with Campbell's reply. He had copied the email to Thomas Cook's own flight ops inspector, along with one other individual. His manager, David Russell.

Although this was to be handled confidentially, I was genuinely concerned that this email to the CAA would find its way back to Thomas Cook via their inspector, despite it being a confidential document. There again, if Thomas Cook knew things were getting serious, this might cause Burke, Hutchings and Scadeng to question whether they were doing the right thing. At least the CAA now knew what was going on.

I replied to David on the same day.

"Thanks for getting back to me.

I'm genuinely pleased that you feel that the information warrants an investigation into the handling of the matter."

I also felt it was important to let the CAA know that I had tried to settle this situation before it had escalated via the email to Scadeng;

"Please accept I have contacted the CAA as a 'last resort'. As an eternal optimist, I had hoped, having reviewed the information over four months, the senior management at the airline would have realised that I did exactly the right thing."

At least it looked like the regulator was taking the issue seriously even if the BALPA CC were not.

There was, however, with regards to the CC, a glimmer of hope on the horizon. Word was spreading amongst the pilot population about how the Scadeng posse on the BALPA CC were prepared to let me hang. It was slowly dawning on them; if it could happen to me, it could happen to them.

The light at the end of the tunnel came in the form of two individuals. Both, I had been informed, had resigned from the CC in disgust at Scadeng's previous alleged antics. There was to be an election to determine who would represent the pilots on the new BALPA CC. Pete and Steve, the two individuals who had resigned, witnessed how the union was now failing its membership and decided to throw their hats back into the ring. They had both stood for re-election.

Within a couple of weeks, the results were announced.

The strength of feeling of the pilot community over the recent toothless responses by BALPA to various issues, including my own, was palpable. A landslide re-elected both Steve and Pete. Steve was elected chairman, Pete vice-chairman.

The difference this made to my case was immediate. Within a day of being re-elected, Pete, who also happened to be the most senior pilot in the entire company and the most senior trainer and examiner, rang me. He wasted no time in getting to the point. He explained that both he and Steve had been appalled at what had gone on. He agreed with my conclusion that this matter was entirely the doing of Scadeng and Hutchings, and Burke was nothing more than a useful scapegoat. It was just she appeared not to be able to see what everyone else could.

Pete informed me that I now had the unanimous backing of the Thomas Cook BALPA CC. I mentioned to Pete how strange it was that as soon as Steve and himself rejoined, Scadeng's remaining old cronies switched sides and fell into line. That told me nothing that I hadn't already worked out earlier.

Pete saved the best until last. He informed me that due to the seriousness of the accusations, the CC had unanimously supported the motion for me to have a full-time BALPA HQ official to now handle the case. Pete put it more succinctly. 'Time to bring out the big guns.'

Steve and Pete had fully realised the precedent it would set if Thomas Cook were allowed to get away with this outrageous behaviour.

The individual at BALPA HQ that Pete named was Dave Margrie.

Margrie's speciality was employment law. Months after the case had reared its head, I was finally getting the level of support that not only I, but my colleagues, would have expected to have received at the beginning of this sorry episode.

A date was set to travel down to Heathrow to meet Margrie and provide him with copies of all the documentation. This was the breakthrough I had been waiting for. With my determination and knowledge of the case and the legal backup provided by the union, Thomas Cook was finally going to get their collective comeuppance.

It looked like my persistence was finally paying dividends in securing the full legal support from BALPA with the CAA now furnished with all of the damning information. Both organisations could now see precisely what had been going on.

It was also time to get my very pink folder off to Scadeng, along with its covering letter. Needless to say the disciplinary on 29 August never occurred.

It looked like I better brush off my best suit and polish the Doc Martens. It was time to visit BALPA HQ at Heathrow!

BALPA HQ lays just off the main road that runs parallel to the main runway at London Heathrow. Theirs is a large building that, for some reason, reminded me of a beehive.

Once inside the hive, there were a significant number of worker bees looking incredibly busy.

I was met in reception by the one member of the CC who had supported me from the start. Richard proceeded to lead the way up to Margrie's office. As we climbed the flight of stairs, Richard explained that as well as Margrie and himself, the head of the BALPA organisation, John Stembridge King, would also join us along with Ian, the flight time limitation specialist. It seemed that once 'Big BALPA' had been given all the info by the CC without the apparent spin imparted by Scadeng, they had immediately given it the attention it should have received at the start of this calamity.

Margrie was sitting at his desk. He stood up to greet us.

Margrie stuck out his hand to shake mine.

'Pete, nice to meet you, take a seat.'

'Mike,' I replied.

'Oh err yes err Mike err, hold on.'

'Now this was about, err er.' Margrie looked at the assembled throng for guidance.

It was clear that maybe the attention I thought I was going to get had been a little presumptive on my part.

'It's the refusal to fly... Mike's the fatigued guy,' prompted Richard adding 'the problem with Mike is he's a shit magnet'.

I snapped back at Richard,

'funny that I have only become a shit magnet since BALPA allowed Scadeng to take the position of chief pilot without gardening leave, isn't it?'

I began to explain to Margrie that this whole sorry state of affairs started only a week after I was contacted by Scadeng to alter the witness statement relating to the taxiing incident that happened four years previously. I was a little surprised when Margrie recommended I did not bring in the connection to this witness statement. In his view, it would '*complicate matters'*.

Margrie began to pick the bones out of the situation.

'Right. You refused to go into discretion?'

He read a bit further.

'I'm confused -- how can it be a management instruction to use YOUR discretion?'

I didn't need to answer, the look on my face did that for me.

'And how would they know you were being dishonest about being fatigued?'

It appeared BALPA and myself were at last singing from the same hymn sheet.

I also told Margrie about the medical emergency and the ramifications of which I had to deal with at the same time as this ridiculous run of duties. I said that Burke dismissed it. She didn't want to know. Margrie's view was that if it didn't fit Thomas Cook's narrative, then they would not be interested.

'You are aware this is a comparatively cheap way of getting rid of you, don't you?' Margrie was about to unload his years of experiences and legal expertise into the case.

He explained that he had seen similar situations before where individuals had raised concerns with their employers relating to legal and safety failures. Instead of dealing with the issues, especially if there was a concern that others might become encouraged to make a stand, the easiest and cheapest way to deal with the situation was to get rid of the whistleblower. It was especially true if the improvements required to counter the safety failures would hit the bottom line.

There was little doubt in my mind this was exactly what Thomas Cook was up to, but '*cheap*'?

Margrie explained that it was vital that I was fully aware that, even if I was dismissed and it was deemed unfair by a judge, the most I could expect in damages was capped at around £72K, or one year's salary, whichever was the lower amount.

Margrie added that the only time there would be no upper limit for damages awarded was if a case was brought against the employer for sexual or racial discrimination.

There was consternation all round at how unfair this was, not least from me! Still, Margrie explained this was the route organisations regularly took when they wanted to rid themselves of high earners who had 'blown the whistle'.

I explained this was never about financial concerns; it always had been about just doing the right thing. There would also be little chance of ever getting a job in aviation again. All because I refused to break the law and put passengers' lives in danger. Is this really how it works?

Margrie continued to advise why I was in such a bad position. He explained all the company would have to do was say they had a valid reason to come to the conclusions that they had. It would then be up to me, in front of a judge, to prove that they did not. In effect, I would have to determine what somebody was thinking. Margrie believed that, in a case like this, the claimant rarely won. The odds were most definitely stacked against me.

The head of BALPA, John Stembridge King added:

'As you see, that's a very high bar for you to overcome, Mike.'

This was bad, really bad, but then Margrie gave a glimmer of hope. He explained that it was possible Scadeng, in his new chief pilot role, had overstepped the mark in an attempt to impress Hutchings. Margrie could see a situation where Scadeng would hold the disciplinary, dismiss or demote me, or issue some other serious sanction. I would, of course, appeal. At which point, Hutchings would then reduce the penalty and demonstrate, in one fell swoop, he was the boss and had the power to override Scadeng. It would serve to not only put Scadeng in his place, but also still provide me with a firm rap over the knuckles for daring to question and refuse to comply with the 'reasonable management instruction'. With my tail between my legs, it would serve to inform the entire pilot and cabin crew community that to challenge Hutchings was just not worth it. It was a terrible career move.

It appeared my family's entire financial security was based on hoping Hutchings would reduce any sanction applied for doing exactly what I was required to do as an airline captain.

Sod that!

BALPA's main selling point to its members is the provision of legal support for any dispute with your employer. I asked the obvious question given the circumstances.

'When can I discuss this with a solicitor and provide them with the details?'

Both Margrie's and Stembridge King's response was the same. They informed me that until Thomas Cook took sanction against me, they had no case to answer. They had done nothing wrong yet.

Nothing wrong?! I explained their actions against me were already causing passengers to be flown by pilots so tired they were falling asleep at the controls. As far as the company were concerned, this level of fatigue was nothing out of the ordinary.

I suggested that if Thomas Cook were made aware that we were in discussion with a legal expert, they would be more likely to drop this ridiculous case. But no, Margrie and Stembridge King were adamant. BALPA would not provide funding for legal advice until after the disciplinary outcome.

This was not what I wanted to hear, but at least I had a better understanding of exactly what I was up against.

There was just one final item on the agenda. I informed Margrie that I had been in contact with the CAA and that I had provided a whistleblowing pack for them on precisely what was going on. I asked him to have a look at the pack to make sure I had not overstepped any issue that would allow Thomas Cook to subsequently use against me. Yes, it was supposedly confidential, but I was not confident it would remain so.

Immediately Stembridge King opined that I had to be *'careful'* not to *'antagonise them'* and cause any more problems.

No, I could no longer hold in my response to such a weak position for the head of BALPA to take.

'John, I have been labelled dishonest and face a disciplinary, I really don't give a crap if it antagonises them or not.'

At that, the meeting ended with Dave Margrie agreeing to look over the whistleblowing documents and confirming both he and Ian would attend the disciplinary that was now set in a couple of weeks' time.

Even though BALPA's response was not exactly setting the world alight, at least the CAA was going to investigate, this despite

Stembridge Kings's reservations.

# CHAPTER TWENTY

## *The Cosy World Of Regulatory Capture*

*"ANO 146 (1) 'A person must not act as a member of the crew of an aircraft to which this article applies if they know, or suspect, that they are suffering from, or having regard to the circumstances of the flight to be undertaken are likely to suffer from, such fatigue as may endanger the safety of the aircraft or of its occupants."*

UK Aviation Law

The date was set for the disciplinary. It was still a couple of weeks away, so the only decision now was when to unite Scadeng with the pink torpedo that was going to sink Thomas Cook's argument. Seeing as though they had given me less than 12 hours to study their investigation folder, it would be rude of me not to return the short notice favour. As an understanding individual, I sent the package recorded delivery to arrive on the Friday before the meeting. Let's ruin Scadeng's weekend by having to spend it witnessing Burke's sixteen weeks of work torn apart.

A week had gone by since sending the documentation to David Campbell at the CAA. I still had no response from the regulator's representative. I decided to ring him.

For a CAA inspector about to launch an investigation into an open and shut case, Campbell sounded very nervous on the other end of the line. I had assumed we were on the same page.

I asked him if he had a date for the investigation.

'Err mm err well no,' was his spluttering reply. Campbell sounded like he expected me to bite his head off. Apprehensively, he told me that not only had he not commenced the investigation but that he was explicitly directed by senior CAA management not to investigate

Thomas Cook. This was even before the senior manager had seen the evidence.

Was Campbell being leant on from on high? If so, it was time to lean on him from the other side then to keep him straight.

'By who, Thomas Cooks flight ops inspector?'

Campbell immediately denied that it was this individual. I continued to press him.

'Come on, Dave… who told you not to investigate?'

Campbell asked me to phone him the next day. He said that the Thomas Cook inspector would be with him and I would be able to speak to them both.

When I hung up, the situation was becoming crystal clear.

In the last couple of weeks, I'd had the meeting with BALPA where the head of the entire union had warned me off offending the very same individuals who, I was claiming, had broken the law. We had a chief pilot who only a few months ago would've been working closely with the very same head of the union as chairman. It could not possibly be that BALPA saw me as a threat to the cosy BALPA - Thomas Cook relationship, could it?

On the regulatory side, we now have a CAA flight ops inspector, told in no uncertain terms by his manager, irrespective of what the facts were, not to investigate my accusations. We have a director of flight safety at Thomas Cook who had previously been a flight ops inspector for the CAA. It couldn't possibly be that the CAA saw me as a threat to the cosy CAA - Thomas Cook relationship could it?

The next day I rang Campbell. He now sounded more confident. Was this a case of safety in numbers?

He informed me that 'We do not get involved in employment issues'. It very much sounded like Campbell had been pre-briefed what to say. As a fob off, though, it was about as weak as they come. I still had the impression, however, that Campbell was just following orders and this was not the way he wanted to proceed. I had Campbell down as a good guy.

I explained that a breach of the law, which according to the ANO and E.U. Ops, amounted to criminal acts, was most definitely, not an employment issue. I made it clear I was not asking for the CAA to ponder the employment aspect, just the offences that would directly result in passengers flown by pilots asleep at the controls and, at the very least, result in a significant '*reduction in the safety margin*'.

At this point, the Thomas Cook flight ops inspector chimed in;

'Well, it depends on how you interpret the law.'

I was amazed I managed to say the following without an expletive, although the volume doubled.

'You're the regulator. You should know the law! There is no interpretation required.'

I put to him an indisputable fact. No one, but no one, would know how fatigued I actually was, apart from myself. No one would know, including myself, how much more fatigued I would become during a duty. That's why the law concerning fatigue is written as it is. Did I really have to explain the law to the CAA? I did, anyway. I read it out, word for word. I emphasised the four key components, *know, suspect, is suffering* and, *likely to suffer.*

I suggested to the Thomas Cook inspector that there was very little to interpret in the word 'ensure' with regards to the FDP breach. I had gathered from the previous day's conversation that the regulator was not keen to do what they were tasked to do, i.e. regulate. Consequently, I had the applicable laws laid out in front of me. To hammer the *'ensure'* point home, I read out the concise and straightforward legislation emphasising the appropriate line.

'An operator shall ensure that for all its flights, flights are planned to be completed within the allowable flying duty period taking into account the time necessary for pre-flight duties, the flight and turn-around times.'

The less than confident reply from  flight ops inspector was:

'But, but circumstances can change.'

Let's chop that argument off at the knees.

'Show me where in the legislation does the word "circumstances" or "change" appear?'

His argument receded quicker than the tide on Southport beach.

Now to the vinegar stroke of today's conversation. If it wasn't the airlines own CAA flight ops inspector, who was it putting the block on Thomas Cook being hauled over the coals?

Finally, after a lot of prevarication, I got a name. Well, to be precise, before I got the name, I got his title. It was the CAA's flight ops manager. It was the Scarlett Pimpernel, or to be more precise, David Russell.

Russell would prove to be as elusive as the Daniel Defoe character.

Eventually, I extracted the manager's email address from Campbell without having to resort to waterboarding.

I immediately set about writing to Russell to *'ensure'* he was aware of all the details. I would also ask precisely why he had told Campbell not to investigate before he saw the evidence.

Unfortunately, in the aviation world, emails do seem to have a habit of disappearing. It's incredible how it is always the emails that people have to, but don't want to respond to, that always find their way into the recipient's junk folder. Let's head that problem off at the pass. I also sent the documents by recorded delivery.

I had received next to no support from BALPA and so far, only a pitiful response from the CAA. Given the evidence, how could I come to any other conclusion that my employer, once a well-respected airline, was now, in my view, run by certain managers who were either dishonest with little integrity or common sense or, was it, they were just plain incompetent? Did I really want to remain part of this industry? Did I want to spend the next ten years of my life looking over my shoulder and consider every command decision could cost me my job? When you got down to it, did I want my working life controlled by individuals I simply no longer trusted? Individuals who appeared, from my perspective at least, to be more concerned about the size of their bonus than the safety of the passengers. This was not the company I wished to keep.

Could I move to another airline? Well, given the issue with my seniority not being transferable and loss of security and income, that was not an option. Fuck it, why the hell should I?

So what were my options?

I could think of only one.

Fight them all the way, do not give an inch, don't rely on anybody but myself. I had zero choices. Wherever this ended I wanted to be able to look myself in the mirror and be able to say, I did all I could. The thing is, what could I do when facing the entire UK airline establishment?

That evening I watched a documentary about Muhammad Ali.

# CHAPTER TWENTY-ONE

## *Rope A Dope*

*"They call it the rope-a-dope. Well, I'm the dope. Ali just laid on the rope and I, like a dope, kept punching until I got tired."*

George Foreman

The 'rumble in the jungle' between Muhammed Ali and George Foreman was one of the most memorable sporting events ever seen. The fast, super fit, young Foreman had everything going for him. Ali, was, by professional boxing standards, an old man, over the hill, past it. He could, in no way, be expected to stand a chance against a much more powerful and younger opponent. During the fight, Ali was pummelled, round after round. It looked like it was going to be a sad end for a once-great boxing legend. For virtually the entire fight Ali was literally up against the ropes.

But for those who were observing, something was going on. Ali spent every round leaning on the ropes. Not only that, but Ali was goading Foreman. Every few seconds, Ali would pop his head around his guard and whisper something offensive to Foreman as he grabbed him to pull him close. This would then illicit yet another ferocious attack. Ali took it. When it looked like Foreman was slowing down, Ali would bounce off the ropes and do his famous Ali shuffle. The intention was clear. No matter how hard you punch me, I'm still standing. This would then result in yet another pummelling. Ali absorbed everything. Rather, he didn't.

Being the master tactician, Ali had concluded, if he leant back on the ropes, every time he was hit, the force and power of the punch would travel through him, and be transferred to the ropes leaving him relatively unhurt. It was the ring that was absorbing everything.

Eventually, Foreman exhausted himself; he had nothing left to throw, no reserve. The moment Ali sensed this, he came out, all guns blazing, and in a few seconds, Foreman was on the ground. Ali had 'roped a dope.'

It was a similar principle to judo.

Thomas Cook were certainly pushing, but with the lack of legal help from BALPA or CAA assistance, I would not be able to push back with any form of legal firepower. During the internal process, they held all the cards; they were throwing the punches.

My only option when they punched was to *rope a dope*. BALPA had been useless up to this point. Although the union claim to offer legal support, I was starting to doubt it. If my own union were going to be as helpful as a chocolate fireguard, I asked myself which union would definitely put its member's interests first. Which one would put the fear of God into an employer? There could be only one, the Railway and Maritime union, known as the RMT. The head of the union was Bob Crow.

After a few minutes of searching online, I came across a pamphlet issued by the RMT and signed by Bob. It explained, in detail, what to do, if there was ever a situation at work where the safety of passengers or rail worker was in jeopardy. It was exactly what I was looking for and exactly what I needed to know.

It was telling that this two-minute search on the internet had been more useful than the collective efforts of a union based in a beehive at Heathrow. I was paying nearly one hundred pounds a month for supposed legal advice. What I was about to find out in the next few minutes was that the only information BALPA had actually given me was 100% wrong.

The main thrust of the information contained within the RMT pamphlet was that if any company took disciplinary action against an individual for walking away from what that individual honestly perceived to be a dangerous situation, that individual was protected in law. The pamphlet referred to section 47b of the Employment Rights Act 1996.

If a railway union was able to provide this information to its members, why had BALPA not done the same? As far as I could discern with my non-legally trained mind, the Employment Rights Act made my case a slam dunk.

Was this incompetence on the part of BALPA? Did they not want to give the pilots the ammunition to fight cases such as mine? If I had not heard of this legislation before, I wondered how many of my colleagues had. I asked several. Not one was aware of it.

Now, I wonder why that might be?

Boy, did I need an employment lawyer on this one? Sections and subsections of legal documents relating to employment law sounded as dry as Jeddah on a Friday. It was Hobson's choice. If I didn't do the research, no one else would, certainly not BALPA.

Having dealt with the intricacies of music business contracts and following the twists and turns that were placed within, with the specific purpose of duping the unwary, maybe I stood a better chance than most of navigating the legal maze I was just about to enter.

It was apparent that there were two particular aspects of the case.

The first was my alleged safety and legal breaches of legislation by Thomas Cook Airlines. This aspect was the CAA's responsibility. Regulate, investigate and prosecute. A failure to do so would put the public at large in danger.

The actions that were taken against me by Thomas Cook for highlighting these criminal and unsafe breaches of aviation legislation were, in themselves a breach of employment law. At least that's what I had deduced.

I tracked down the relevant section of the Employment Rights Act 1996. Amazingly it was not at all complicated. It revolved around something referred to as 'protected disclosures' (PD's). I had never heard of such things in my entire life. This was a good sign. In fact, it was a great sign. If the experienced airline pilots I had talked to were not aware of PD's, you could be pretty confident Jo Burke and Scadeng would be equally clueless.

A Protected Disclosure can best be described as when an employee discloses information that in his reasonable belief was made in the public interest that tends to show one or more of the following:

(i) That a criminal offence has been committed, is being committed or is likely to be committed;

(ii) That a person has failed, is failing or is likely to fail to comply with any legal obligation to which he is subject; and/or:

(iii) that the health or safety of any individual has been, is being or is likely to be endangered.

The critical point is that an individual must have a reasonable belief that there is a serious issue.

In simplistic terms, a protected disclosure is more widely known as whistleblowing.

The critical aspect of protected disclosures is that it is illegal, for obvious reasons, for any company to take any action against an individual who raised genuine safety concerns. If a company did, then this would be classed as a 'detriment to your employment'. If this did happen, then it was possible to take legal action against your employer.

I had now been suspended for six months. Did this qualify as a detriment to my employment? Again a quick Google search indicated that it did. After all, I had lost thousands of pounds in duty pay.

All the above seemed to confirm that my case was as watertight as a submarine's hatch.

Margrie had already highlighted the only remaining issue. He had stated that even if I won every point in front of a judge, I would still only be awarded a maximum of around six months' salary in damages. My case was not based on sexual or racial discrimination. According to Margrie, these were the only claims that would warrant unlimited damages.

He was wrong.

I discovered anyone who experienced a 'detriment in their employment' due to raising safety issues was in the same position as anyone claiming it was due to sexual or racial discrimination. There was no upper limit in damages.

How could BALPA get something so basic so wrong?

Was it an attempt by BALPA to dissuade me from proceeding with the case? Was it aimed at placing doubt in my mind as to 'was it worth it'?

It left me with a big tactical decision. Do I inform Scadeng during the disciplinary meeting he is staring down the barrel of a major PR disaster and unlimited damages?

While pondering that very question, within thirty minutes, I had two phones calls that made the decision for me.

The first was from Richard from the CC. He said he had talked to Scadeng and that Scadeng had informed him he was going to '*nail me*' with, as Richard put it, the Burke produced '*dodgy dossier*'.

The second call was from Steve, who was now the head of the BALPA CC. He said he had also had a conversation recently with Scadeng where Scadeng had referred to me as dishonest. I asked Steve would he be prepared to confirm what was said in a written statement. My view was that Scadeng was knowingly defaming me and could result in a case for slander. At the very least it would put the frighteners on him. Steve said he was not prepared to do that as 'I've got to work with him, and it was said in confidence'.

Decision made. It was time to rope my own dope.

# CHAPTER TWENTY-TWO

## *Get On With It*

*"I'm trying to understand the difference between fatigue and tiredness."*
Roger Scadeng, Thomas Cook Chief Pilot

The day of the disciplinary finally dawned. I fully expected today would be my last day employed as an airline pilot. One way or another, the next few hours would define my future career within the aviation industry.

I had arranged to meet Ian from the union at an 'all you could eat' breakfast buffet at a hotel located only a few minutes away from the airport. Ian was already tucking into a full English by the time I arrived. I joined him in the cholesterol-infused delicacies of sausage, hash browns, bacon and black pudding.

Between alternating mouthfuls of tea and fried food, Ian remarked:

'For someone facing the sack today, you seem pretty laid back about it. Are you not worried?' Good, Ian was picking up exactly the vibe I wanted to exude.

I shrugged, 'would it help?'

In all honesty, I was totally at ease about the meeting and what might or might not happen. I had my game plan, and I was going to stick to it. I had boxed Scadeng and HR in via my letter and documents in the pink folder. Everything was in there, hard evidence rather than supposition. The letter stated that if Scadeng attempted to fabricate anything, he would be vocally and publicly called out at a later date. What happened long term was no longer within the remit of Scadeng, Hutchings, or any other Thomas Cook manager with an agenda. I was in control. They just had not worked that out.

For my strategy to work, I had to keep my own council. I could not reveal the long term game plan to anyone. Not for one second did I question Ian's integrity, but, like Dave Campbell, he might inadvertently say something to someone that might get back to Thomas Cook. Once the airline worked out what was going on, they would have no other option but to fold. If they didn't, I would rope my dopes. I would put money on they wouldn't. Their arrogance would not allow them to.

Dave Margrie joined us a few minutes later. Both Ian and Margrie were wearing suits, shirt and tie. 'Unusual choice of attire for a disciplinary,' was Ian's verdict on my wardrobe selection. Perfect. If these guys were picking up the contempt I had for the entire process just by what I was wearing, so would Scadeng. Obligatory black Doc Martens, polished, of course, black shirt, no tie, and straight leg black jeans were the order of the day.

We arrived at the Thomas Cook Airlines head office thirty minutes before the meeting was scheduled and headed straight for the canteen. As soon as I sat down, I was approached by numerous well-wishers who had heard the rumours surrounding the case. One comment by a crewing manager who I had known for years summed up the general consensus. 'Give them a kicking. They deserve it, and someone has to.'

Imagine my surprise when we were informed that Scadeng was not 'quite ready'. Utterly predictable. Let him play his little games. There was then a discussion between the three of us about Scadeng's suitability to chair the meeting. It was fairly evident that he should be nowhere near the disciplinary. However, for my long term strategy to work, it was important he was the one conducting it. Whatever Scadeng decided today was immaterial. The important thing was, the more I allowed them to disregard and ignore their own laid down procedures, the better. I was now playing the long game.

As we sat in the cafeteria, Margrie was becoming more agitated by the minute. Scadeng's delay was starting to get to him. As for myself, I was fully expecting this fiasco. I suspected Scadeng had not bothered to do his homework over the weekend and had left it until this morning to look at the pink file. The less care he took in reviewing the documentation, the more likely he would miss the little surprises that I had liberally scattered within its pages that would bloom at the appropriate time. Margrie suddenly announced 'this is seriously pissing me off'. To kill some time he re-read a copy of the covering letter I had sent to Scadeng along with the pink folder. He glanced over the A4 paper. 'It's a little late but don't you think it's a little aggressive?'

To Margrie's mind, it sounded like a veiled threat. It was in no way a veiled threat. It was a gold plated guarantee that if they wanted to play silly buggers, they had completely misjudging not only my determination but also my ability to hold them to account.

Margrie again reminded me that even if I won my case in front of a judge, the cap on damages would have a profound effect on my financial situation.

I just smiled before highlighting the main thrust of the argument to Margrie and Ian. All I needed to do was to tell Scadeng the reason for the disciplinary was because I had raised safety issues. That's it, nothing more, nothing less.

At that very moment, Scadeng walked past the doorway of the canteen with his face the same colour as the bright pink folder under his arm. Behind him scurried three HR elves. They collectively looked extremely stressed. I couldn't help myself, 'what a lovely folder' I remarked.

We had been kept waiting for nearly an hour before we were eventually ushered into the chief pilot's office. Scadeng did not even look up to acknowledge our arrival. The old trick of trying to convey that we were not as important as his other chores was in full effect. 'How To Try And Intimidate People For Dummies' (1972 edition) must have been his reference material. No greeting, no apology for wasting an hour of my life.

Oh boy, was I going to enjoy this.

Today was the very first time I had ever set my eyes on him. He was exactly how I imagined. For some reason, he reminded me of 'Captain Mainwaring' from the TV show 'Dads Army'. I also noticed he was already profusely sweating. If he was sweating now, goodness knows what he would be like by the end of the meeting.

Next to Scadeng sat Harragan. Harragan must have gone up in the world. She was no longer the note-taker. Instead, another lady was sat, pen in hand, to her right. It did beg the question, what exactly was Harragan here for? Was she on brow-mopping duty, ready for the tsunami deluge of sweat, or was she there simply to hold Scadeng's hand? It was, after all, his first disciplinary.

It was time to rope a dope. Sitting down, I immediately leant back on the ropes. To be more exact, I rocked back on my chair and stared. Arms folded and legs crossed with my right Doc Marten boot resting on my left knee. I was goading him to say something. He didn't rise to the bait. He still had not looked up. As I studied this individual, another trickle of sweat began to make its way down Scadeng's forehead.

By the time he realised that I was more interested in the track of perspiration than making any attempt at a greeting, he looked up. After finally making eye contact, he stuck out his hand. I stuck mine out. The distance was about three feet between our respective hands. I was still leaning as far back on my chair as I could without toppling. My Doc Marten boot still resting on my knee. I had no intention of moving forward. I was too busy keeping my balance and enjoying every moment.

He leant forward, and I stayed where I was, arm still outstretched. Closing distance now around two feet. Again a pause. Now Scadeng was leaning over the table having to support himself with the other hand. I shook his hand. It felt like shaking a damp flannel.

Round one to me.

As soon as he sat down, I allowed all four legs of the chair to connect with terra firma.

Scadeng began reading out the now well-worn allegations. He appeared to be savouring every sentence, and every word. Boy did he want to let me know how deep in the crap I was. Like the schoolboy impressing the girls as he rides past on his bike. *'Look at me, no hands.'* I had a distinct feeling this show was as much for Harragan and the note-taker as it was for me. The sideways glances he gave tended to support that conclusion. *Yes, look at you, by the end of the meeting, you will metaphorically 'have no teeth'.*

Margrie's immediate response was to draw attention to the bundle of documents that I had provided in the DayGlo pink folder. Scadeng had pushed the folder as far away from himself as he could without it falling from the table. It was so bright, it almost pulsated. It represented everything that would ruin his argument. Maybe he concluded that the physical distance of the file would be directly proportional to its impact. It was, however, the complete dossier. Within its contents, I had covered off every single allegation. I also had indicated that the only logical conclusion that any reasonable person could possibly come to, given the facts, was that these allegations had been concocted by an individual, or individuals who were either incompetent or of dubious integrity. As long as I kept dragging Scadeng back to its contents, he would fail in whatever he thought he would achieve. It was literally impossible for there to be any other ultimate outcome. To ensure there was absolutely no wriggle room for

Scadeng, I hammered the point home.

'My views and recollections are contained within the bundle, which I require to be taken into evidence.'

Scadeng then proceeded to ask every question that I had already answered in the pink file. His tactic was exactly like Burke's had been at the investigation meeting. When he realised that I had closed off a line of questioning, he would circle back around and ask the exact same questions, but from a different angle. The answer I gave to any particular line of questioning was always the same. In footballing terms, I had parked the bus, and this particular bus was not red, but bright pink.

Another question from Scadeng. This time I would reply by quoting chapter and verse the legal documents, the Air Navigation Order and EU, Ops. It made not one bit of difference to Scadeng. He was a man on a mission, and the facts and legislation were a mere inconvenience.

As the disciplinary continued, Scadeng became visibly more and more agitated with his lack of progress. Every avenue he tried to go down was met by a dead end, a metaphorical brick wall.

Scadeng's complexion had also changed to the short wave of the spectrum. It looked like the rope a dope tactic was starting to bear fruit.

Scadeng snapped 'did you believe the duty was rostered illegally?'

If he had read the bundle, he would have known the answer.

My answer, like all my answers, was taken, virtually word for word from the folder. No opportunity for Scadeng to weasel out any comments that I made, or to then take them out of context. They were already there, in black and white. This was why I wrote the covering letter. It was to strangle at birth any ability for Scadeng to pervert this disciplinary meeting.

Scadeng immediately changed tack, realising he was not going to get any purchase on that front.

'Do you think it was reasonable to assess your fitness for the following day?'

The fact that the chief pilot would even ask such a question showed how desperate the whole situation had become for him. At this rate, not only would I rope a dope, I would have him trussed up with an apple in his mouth by the end of the meeting.

The correct answer to Scadeng's question was, not only was it reasonable to inform the company but, Thomas Cook's own manual required me to "assess my fitness for the following day". The company documentation demanded it, and if I didn't, I could face a disciplinary.

Yet here we were, a desperate chief pilot asking me did I think it was appropriate to fully conform to Thomas Cook laid down procedures that Scadeng himself was responsible for issuing and keeping up to date.

At this point, nothing would have given me greater pleasure than to lay into Scadeng and let him know what I thought of him and his attempted stitch-up. No doubt it would have felt magnificent, at least for a short time. This, however, was not in the plan. Delay the pleasure, and just grab more inconsistencies to embarrass him at a later date. Delayed satisfaction is always worth waiting for.

So do I highlight at this point this fundamental flaw in his question?

Of course not. Let him wander further into the spider's web. Again leaning back on the chair with legs crossed, I answered his question. 'In order to protect the company, yes.' This was my intentionally limited response.

I wondered how long it would be. Scadeng shuffled in his seat, the hint of a smug look across his features. I could see it coming a mile away. Scadeng was about to reveal his checkmate moves that would result in what had been intended all along. It would give him and Thomas Cook the reason to take disciplinary action and to dismiss me on the spot. Scadeng was becoming more and more excited at the prospect of his next statement. He sat back and started to push his papers around. It was apparent he wanted to savour every last moment of my impending discomfort and observe how I would react witnessing this masterful tactician in full flow. He then leant forward and leaned on the table so I would experience the full impact. He had his serious face on.

'Were you at any stage in the last sector too tired to go into discretion?'

'No.'

It was as clear as daylight what he was trying to do. I was five moves ahead.

'So you were not at any point too tired to exercise commander's discretion.'

'Correct.'

So there was another example of asking the same question twice and Scadeng getting the same answer.

Finally, after months of preparation, Scadeng unleashed his carefully crafted onslaught.

'You can be fit enough to operate beyond max FDP yet at the same time sufficiently fatigued not to operate beyond minimum rest. How is that possible?'

Was that it?

The silence suggested it was. That was all he had.

The thing was, he didn't even have that.

Scadeng's somewhat disjointed question was central to his apparent attempt to fabricate a case against me.

The recorded meeting with Burke and her fixation on the HF conversation had transmitted clearer than any memo could, what Scadeng was up to.

In order to claim I was dishonest Scadeng had to establish, somehow, that during the last sector, I had claimed that I was fatigued. His closing gambit would then be to declare that I could not possibly have been fatigued because I had agreed to operate into discretion. Therefore I must have lied about being fatigued in the first place. Job done, convoluted, most certainly. It made Leo Tolstoy's War and Peace sound like Cat in the Hat. His theory and grasp on reality, however, was about as tenuous as this analogy and made as much sense as dr Zeus.

The thing was, and this was precisely Scadeng's and Thomas Cook's big problem: at no point during the HF transmission had I claimed I was fatigued during the flight. Indeed at no point had I ever claimed I was fatigued that day, not on the HF call, the phone call, the meeting with Burke or in any of the paperwork. Scadeng had spent the last several months attempting to smash this square peg into a round hole. He had hold of a piece of a jigsaw that just did not fit. Scadeng's dilemma now was he had to, somehow, complete the puzzle. He somehow needed to make the piece fit for the ex-BALPA CC chairman not to lose face in front of his new master. If the piece doesn't fit, what do you do? You huff and puff, you bang it with your fists, you stamp on it in an attempt to distort the entire puzzle until the picture bears no relation to reality.

I looked at the desperation etched on Scadeng's face. Did he know he was drinking at the last chance saloon? His only hope was for me to state I was fatigued on the final leg and say it during this meeting.

The last vestiges of any hint of his smug demeanour had now been washed away by constant perspiration. He was about to hit the brick wall of his argument so hard his teeth rattled.

I final gave Scadeng his answer and read the letters out slowly, each one emphasised.

'F.R.M.S.'

From the blank look on Scadeng's face, I might as well have been speaking Mandarin.

I was referring to the companies FRMS (Fatigue Risk Management System) manual that every employee was required to read and understand. It went into great detail about fatigue and its causes. How long duties, shift changes, long haul and such affect the circadian rhythm and sleep pattern. It's basic common sense. Anyone who had even the slightest perception of the principle behind FRMS would now see how deep the hole Scadeng was now in and therefore how absurd his question was.

No guesses for the name of the individual to who was, according to the operations manual, responsible for ensuring the FRMS manual was correct and up to date?

In my carefully considered opinion, given what I had seen and heard over the last few months, this guy was a walking disaster and likely to cause a company who could little afford it, a whole heap of bad press. As I stared back at him, it struck me that even his name sounded like a plane crash.

The manual Scadeng was required to have read and understood, and keep fully up to date, would have answered his question if he had only bothered to read it.

Was I going to do Scadeng's homework for him? Was I hell. Was I going to make a big song and dance about what the FRMS manual actually stated? Not a chance.

My stonewall defence was starting to irk Scadeng. My only problem was my right leg was now getting a little numb. As I watched him frantically shuffling papers, I now shuffled my legs.

I considered the position he was now in. As a drowning man grabs at anything in a frantic attempt to stay afloat, Scadeng appeared now to be grasping at straws. If the straw didn't exist, then he was going to have to invent one—the original straw man.

As far as I could now see, Scadeng had only one option.

All he had left was to try and engineer a false narrative. This would be his only way of supporting any allegation that I was dishonest. To do that, however, he himself would have to 'lie through his teeth'. Lying during an internal investigation is bad enough. The question was, would Scadeng be foolish enough to begin walking a path that ultimately would require him to attempt to mislead a judge?

I had boxed him in completely. If he was going to be stupid enough to lie about when I claimed I was fatigued, he faced a major problem. None of the documents would support this falsified claim.

I suspected Scadeng himself was now starting to understand the predicament he was in. He changed tack.

He questioned why I had not gone into details during the noisy HF call about why I was refusing the following days duty.

'Because it was agreed I would talk to crewing when I landed.'

Scadeng was getting more aggressive, a combination of the brick wall he was facing and the fact that I was still leaning back on the chair with not an apparent care in the world seemed to be working like a charm. He was now overreaching.

'But why not explain on HF?'

His attacks were becoming ever more flaccid.

It was time to come off the ropes and gee him up with an Ali shuffle aimed at tormenting him.

'Roger, how many interceptions has the company suffered this year?'

I was pouring salt into a particularly raw wound for the company. It was a direct reference to the number of Thomas Cook aircraft that had been intercepted by military aircraft over the past year. It was significantly higher than any other UK airline. Since 9/11 the authorities had become extremely nervous, for obvious reasons, when commercial aircraft lost contact with air traffic control. The standard procedure when communication could not be re-established was to scramble a military jet that would form up on the unresponsive airliner's port wing. Ultimately, if communication could not be re-established, a decision would have to be made with the possibility of the plane being shot down. It was that serious. Usually, both the captain and first officer monitor the relevant air traffic frequency. This is to ensure if one pilot misses a transmission, the other pilot can intervene. If the other pilot is on another frequency, the more likely calls will be missed. It was for that reason radio calls that were not ATC, or safety-related were kept to the absolute minimum. Indeed this was so important that during the six-monthly simulator checks, training captains would attempt to distract the crew with non-vital radio calls. If you allowed yourself to be sidetracked, it would result in a black mark.

Given the above, how would Scadeng respond to such a barbed, direct question that required a simple number?

He didn’t, he just ignored it

'Do you believe that if crew are not too tired, the company have a realistic expectation that crews will operate into discretion?'

Another opportunity for another jab:

'Your belief, my belief is neither here nor there.'

Waiting, waiting, there it is, Scadeng was starting to blow himself out, starting to wilt.

Another nonsensical question followed.

'Do you need to be fatigued to operate into discretion?'

That is the equivalent of asking an HGV driver does he need to be drunk to drive. Scadeng, by the minute, was losing it. I had not even hit him with the haymaker yet.

OK, it was time to bounce off the ropes.

I had very gently nudged Scadeng into range without him even being aware of what was happening. Now it was time. The information I was about to reveal would have indicated to any company, or individual, with even the slightest sense of self-preservation or knowledge of employment law that it was now time to wave the white flag. As I had hinted in my covering letter to Scadeng, any inappropriate decision by him would result in dire consequences for not only individuals but also Thomas Cook. The time had come to wheel out the big guns and set the foundation for the long game.

I was confident both Scadeng and Harragan would have no idea what was coming. Even if they did, by now there was little they could do about it.

I began by informing the note-taker that I would read a prepared script. I read the first line.

'The following needs to be minuted exactly for future reference.'

The HR note-taker nodded in acknowledgement.

I then read out precisely what had happened, in fine detail and what both my and Thomas Cook's legal responsibilities were. It was intentionally detailed and intentionally long. The aim was to hide within the pre-prepared script a phrase that I suspected would go totally unnoticed by everyone there. It sounded innocuous; it was, however, as far as this matter was concerned, my 'fat man'.

I was speaking in a slow, laboured, tone as the note-taker was struggling to keep up. It was precisely what I wanted to happen. It forced Scadeng to remain quiet. I knew it would wind this individual up in that I now had control of the meeting he was supposedly chairing. I continued slowly. I emphasised the moment I informed the company I was fatigued. I emphasised the date.

'As I stated in my email of the 7th May to the chief pilot, this was due to my "very real concern that I was experiencing the onset of fatigue which would erode flight safety if I completed today's duty."'

By this point, Scadeng had changed colour from red, all the way back to the short wave end of the spectrum. He was now an aneurysm inducing purple. Would he last the meeting? He could resist no longer.

He bit.

Finally.

He demanded to know when I told Thomas Cook I was fatigued.

'On 7th May.'

I continued my pre-prepared long-winded exposé of the events surrounding the case. Finally, I was getting to *the* moment. Would they react?

'I brought to Thomas Cook's attention, by reasonable means, i.e. an email sent to yourself, and fatigue report, the circumstances re work which I thought would be harmful or potentially harmful to the health and safety of myself, the crew and the passengers.'

Only time would tell if Scadeng or Harragan had picked up on the significance of the text I had just sneaked in under the camouflage of the verbiage of legislation and my views on safety. It was the booby trap that would explode in his, and the company's, face at the appropriate time.

His face indicated he was clueless.

What Scadeng did realise was that he had lost control of his carefully-orchestrated disciplinary. Was it starting to dawn on him that he did not have what he needed to complete his master plan? Did he know the tables had now been well and truly turned?

As I laboriously read out my prepared script, Scadeng's continual attempts to interrupt, snipe and nitpick was a testament to the power shift within the room. Even Harragan was, by now, giving him sideways looks. Had she now realised she was on the wrong side of the argument? After the umpteenth interruption, it was time for me to hammer another nail into Scadeng's ego. I halted, looked up and paused for maximum effect.

'Roger, do you want to know the facts, or are you going to continually interrupt me?'

Right there, right then, Scadeng *lost it*. It was a wonderful moment to behold. He began waving his arms about so much he could have been auditioning for the role of Ian Curtis in the tribute band Joyless Division.

His hands flapped about in front of my face. I suspect he thought this was intimidating. It was a hilariously sweaty sight.

With the greatest amount of contempt he could muster, Scadeng barked 'Get on with it!'

It reminded me of seeing a toddler throwing a tantrum in a supermarket. Harragan looked like she wanted to slap his legs.

This was all my Christmases come at once.

I slowly put the script on the table and allowed my glasses to slip down my nose for maximum effect. It instantly created the headmaster vibe, a headmaster who was just about to admonish an errant schoolboy.

'Roger,' I paused and looked over my glasses, 'did you just tell me to get on with it?'

From the shocked expression on his face, it was clear he recognised he had overstepped the mark. He looked like he was expecting me to jump up onto the table at any moment and bury him in a flurry of flailing fists. I, however, considered that would be taking the Ali boxing metaphor just a little too far.

'No, I said carry on,' was Scadeng's feeble response.

'No, you didn't. You told me to get on with it.' Again I paused.

'Have you quite finished?'

Those few seconds displayed precisely the mentality of the individual sat opposite me. It was pure Scadeng.

If Scadeng was willing to deny what he had said to my face less than twenty seconds ago, what else would he be prepared to claim?

I continued until Harragan intervened;

'Sarah is having difficulty keeping up with what you're saying.'

I bet she was... I bet she was.

Harragan wanted to close this line of attack down and quick, but it was too late. It was game over. I had explicitly injected what I needed to mainline into these proceedings thanks to the Bob Crow leaflet. It was kryptonite to Thomas Cook's position.

Margrie suggested, 'Maybe you could provide the company with a copy of what you're saying'.

Slowly I put the paper down onto the desk and took off my reading glasses. Looking up, I was met by Scadeng's eyes boring into me. I just stared back. It took quite a bit of self-control not to give him a cheeky wink and a smile. Neither of us blinked, neither of us smiled. Neither of us did anything other than just stare back.

About fifteen seconds passed. Through gritted teeth, Scadeng finally muttered,

'Have you finished?'

I could not resist it. I looked at my watch.

'I finished about 15 seconds ago.'

Now he was as wound up as my watch.

He blundered on, and on, and on. Still trying to mine the empty seam of when I claimed I was fatigued.

'At the point, you informed the company you would be operating into discretion did [you] mention fatigue?'

'No, because I agreed I'd call crewing when I landed'.

Harragan was desperate to bring this round to a close. Had she spied the bomblet I had placed into the proceedings earlier on? Maybe she wanted to highlight the collateral damage to Scadeng and the company if this was allowed to detonate. She suggested to Scadeng this would be an ideal time to take a break. With that, Scadeng got up and stormed out of the room with the door slamming behind him. Harragan and the note-taker trailed in his wake.

It was at this point Margrie announced that in his years of being a union rep he had never seen anything to match what he had witnessed for the last hour or so. Unprofessional, biased, with zero interest in considering the evidence. That Scadeng had asked so many questions, already answered within the DayGlo folder, bore testament to that. Margrie seriously questioned if Scadeng had received training on how to either conduct or indeed behave at a disciplinary.

Eventually, the merry band returned to the office. We were also now joined by Ian White.

It didn't take long for Scadeng to once again get over-excited. White had been asked by Margrie to comment on the illegal duty.

'Looking at the FDP allowable at .82 would have been 12hrs 30mins, too tight. There should always be a buffer because if planned to the absolute limit, you'll always need to use discretion.'

Off Scadeng went, like a rat up a drainpipe, literally yelling 'Write that down! Write that down' at the poor note-taker. Again, he was waving his arms around while jabbing his finger in her direction.

'Ian White has just said all flights should be planned with a buffer on the FDP; otherwise, you will have to use discretion.'

Margrie, White and even Harragan exchanged looks at yet another outburst from a man who not only was not in control of the meeting anymore; he had lost control of himself. Harragan suggested that everyone calmed down. I don't think I could have been any calmer without being horizontal. I was enjoying the performance. I was witnessing an aviation version of the Jeremy Kyle show.

Ian immediately corrected him.

'I didn't say that.'

Just to put another, well-polished Doc Marten boot in, I added:

'The flight wasn't planned on the limit. It was planned over.'

Ian then raised the issues of the FAST printout. Thomas Cook were attempting to use the fact that the trace hadn't dipped into the red to claim I wasn't fatigued.

'Do you think we've used it inappropriately?' Scadeng countered.

Ian confirmed it should not be used within the disciplinary. But, here we were, and Scadeng thought this was the evidence he needed to support Thomas Cook's claim.

My position was simple. I didn't need a piece of software to tell me if I was too knackered to fly an aeroplane safely.

At this point, Margrie piped up to further question Thomas Cook's use of the FAST printout.

'I wasn't clear why it was in the bundle.'

'Why would the company suppress any available document?' was Scadeng's immediate and unguarded reply.

Why indeed, Roger? Why would the company suppress available documentation? Why had the document that proved you had been interviewed not been included in the bundle? Why had the company suppressed the flight plans that would conclusively prove, once and for all, that Thomas Cook had falsified that the duty was able to be completed within the legal maximum?

The highlighting of this wilful withholding of documents would have much more impact in the future. Instead, today, I would just give Scadeng a tickle that I was fully aware evidence was being suppressed.

It was an opportunity that was too good to waste, and I took it.

I asked him why he had still not replied to the question I had asked him in the still unanswered email. How was the duty planned that he claimed was legal?

'Commercial produces a plan,' was Scadeng's first attempt to deflect the question, while making no attempt to answer it.

I continued:

'It hasn't been provided to me, and I asked for it over five months ago. So, what speeds did the company use if even at Mach .82 the flight would still not have returned to Manchester within the max permitted FDP?'

Scadeng's reaction to that simple question personified not just his, but all those involved apparent contempt for both myself and any concern to be seen to hold a fair and equitable disciplinary process. Scadeng looked at me, smirked, then shrugged his shoulders.

That was his well-thought-out reply. He didn't have an explanation.

I knew as I looked at this character, he had no idea what was coming. I would have to wait to see him smirking the other side of his face when he was stood on the witness stand.

His next question was beyond parody. For someone who was actively trying to destroy my career by claiming I was being dishonest about being fatigued, he seemed not to understand the basics about the issue.

'I'm trying to understand the difference between fatigue and tiredness.'

*Yes, Roger, and I am trying to understand how you are a chief pilot.*

Again, it would have been easy to provide Scadeng with the definitive ICAO definition of fatigue. A definition of which, due to his legal responsibility, he should have been able to recite chapter and verse. But why would I? He had already made his mind up, and why would I loosen the noose that he had not noticed tightening around his neck?

I had all I needed. It was time to finish him off.

Jab.

'We established that only I can say if I'm fatigued or suspect fatigue. No one in the company can medically qualify fatigued or not. The letter says I was dishonest about declaring fatigue. In order to make that statement, I would have expected the company to have taken advice from some sort of medical expert. There needs to be proof in order to make that statement.'

The unsurprising response from Scadeng:

'There is no requirement for proof beyond a reasonable doubt in a disciplinary hearing.'

Another haymaker inbound.

'OK Roger, where's the reasonable doubt?'

Visibly flustered yet again by a simple question Scadeng spluttered, his temples pulsing. At this point, I considered it 'evens' that he would be carried out of this room on a stretcher.

Scadeng spat out 'Not here to answer your questions. Here to establish the facts of the case'.

Facts of the case? That was about the only thing it appeared,

Scadeng was not interested in. Scadeng had also still not realised, due to his failure to answer the questions I put to him months ago, he was now in the front seat of what was going to be a spectacular car crash.

Margrie closed the meeting. He hit the proverbial nail on the head. He waited for Scadeng to calm down.

'There is no evidence to refute. Furthermore, I have been doing this job now for a long time, and I have attended many disciplinaries and, it is rare for me to say something like this but Roger, there is no case to answer here.'

To this day, I am convinced that Scadeng had set up the meeting as a show trial and fully intended to dismiss me with immediate effect. Was the only reason he was foiled because Harragan had seen the protected disclosure ambush I had set? Did she inform Scadeng of the immense financial repercussion due to there being no upper limit in health and safety cases during the break?

To underline this corporate back-peddling, when asked by Margrie how long it would be before a decision was made, Scadeng replied:

'To be fair, this has taken some time today, and I think it would be unreasonable to rush. I think it's appropriate to take an adjournment and review everything before making a decision.'

Job done. Scadeng was bottling it.

This was the last chance for Thomas Cook to row back before the respective legal hounds from the CAA and BALPA were let loose. Both organisations would surely rip the company to shreds in what had to be one of the most transparent cases of abuse of fatigue and flight time legislation anyone could imagine.

# CHAPTER TWENTY-THREE

## *This Raises Very Serious Concerns*

*"The Dunning-Kruger effect is a type of cognitive bias in which people believe that they are smarter and more capable than they really are. Essentially, low ability people do not possess the skills needed to recognise their own incompetence."*

Dunning-Kruger

Several weeks pass. No minutes from the meeting that were promised, no reply from Scadeng to my email.

Was this due to incompetence or had the penny finally dropped that they were screwed? Who knew anymore? More to the point, I no longer cared. I was confident I knew where this matter ultimately would conclude, and it would not be at the hands of individuals I would not trust to wash my car, never mind run an airline.

We were now in the 'phoney war' period. I had nothing to do with the company, and they had nothing to do with me. My colleagues, however, were starting to pass on information about other examples of Thomas Cook blatantly ignoring FDP legislation designed to protect the public and crews' safety.

I was informed of a case that happened only a few weeks after I had been suspended. It occurred on a Thomas Cook flight from Manchester to Orlando. The trip was to return with the same crew into Glasgow. The company then required the Airbus 330 back into Manchester ASAP to commence another long haul flight. Since the legal limit on-duty hours would prevent the same crew flying the aircraft back to Manchester, Thomas Cook was required to have a fresh crew ready to pick the aircraft up when it landed in Glasgow. As in my case, that was not a choice. It was a legal requirement with which Thomas Cook must comply. If they didn't, yet again, it would be a criminal act.

As the aircraft established in the cruise outbound, the captain informed the first officer that Thomas Cook had contacted him a few days previously. Crewing had stated they did not have the crew to cover the last sector from Glasgow to Manchester. They, therefore, asked if the captain would to agree to fly the last sector back into Manchester. Crewing had guaranteed it would be legal to do so.

This was sounding like such an ever-familiar story.

The first officer expressed concern that both he and the captain would not legally be able to fly the last sector. With that, they both checked the documentation to find that the first officer was right to question the duty. It was a criminal act to plan for the crew to operate the last sector in that it breached the maximum legal limit by over two hours and forty-five minutes.

Exactly as in my case, there were no technical issues. They were discussing this matter the day before the proposed flight and nearly 40 hours before the planned landing into Glasgow. The use of discretion to cover a criminally-planned duty was, therefore, also illegal. Both pilots agreed that they would therefore refuse to operate the last sector.

The next day at around midday, before the report, the first officer downloaded the flight plan for the return sector but noticed that the entire flight had been planned at the maximum cruise speed.

This again sounded very familiar.

There was also a request to tanker fuel into Glasgow to save time on refuelling and that no cleaning or de-catering be provided in Glasgow to achieve a quick turnaround.

To find out exactly what was going on the first officer rang the captain. According to the documents I was provided with, the first officer stated that the captain informed him he could "force me into three hours of discretion". This is not the case, and it is the responsibility of any pilot or cabin crew to refuse to operate if it is either illegal or compromises safety.

The first officer, due to the pressure exerted by the captain, agreed to operate the illegal duty.

Although the first officer ultimately bowed to the pressure placed upon him, he was so concerned by the actions of both Thomas Cook and the captain that he filled in a Mandatory Occurrence Report Form (MOR) that was sent to the company's safety department. The questions he asked were very similar to the ones I had asked the company and the regulator. He expressed a genuine concern of the "abandonment of company SOP's in order to facilitate a quick turnaround". He also highlighted the extra cost in carrying the extra fuel from Orlando through to Manchester. This cost the airline thousands in additional fuel burn to carry the extra six and a half tonnes of fuel.

The final comment in his report was damning.

"The current lack of crews is, intentionally or otherwise, leading people to make decisions that have the potential to impact on flight safety. The captain's decision to operate the final sector was laudable, but in my view, it was outside his authority to exercise discretion. An alternative plan was available that he elected not to use for his own personal reasons. In Human Factors they talk of Red Flags, this particular sector resembled the Kremlin and Red Square on a May Day bank holiday."

So, what could be the possible reasons for the captain to operate illegally?

Was it because he did not want to risk career jeopardy by refusing, like I had, a 'reasonable management instruction' ?

Was it *any* wonder that the company wanted to make such an example of me? It was a clear threat to anyone else who dared put safety before profit and anyone who took any action that would threaten the director's bonus. This intimidation seemed to be working.

Even at this late stage, Thomas Cook could still extricate itself from this mess. I didn't, for a moment, expect that they would. Everyone within the company now knew about the stand-off. It would be a massive loss of face if they now climbed down. The general feeling amongst my colleague? Hutchings and Scadeng were incapable of it.

More importantly, by this point, I didn't want them to. I was ready for the fight. They had made the crucial mistake of attempting to bully and intimidate probably the last person within the entire company who would roll over and accept it. Not my words, but those of pretty much every one of my colleagues who I spoke to.

I had still heard nothing, and it was a miserable day. The weather had changed for the worse with the summer now behind us. The lashing rain made any outdoor pursuits a washout. The incessant checking, then rechecking of all the documents relating to the case had become an addiction, a natural reaction. Fight or flight, literally. Even though the disciplinary aspect had concluded, I instinctively knew this was just the start. Any time I was not dealing with the documents relating to the case, I would be searching for legal examples that might prove useful further down the line. I badly needed a break.

I decided to watch a DVD. The film of choice was 24 Hour Party People, a movie starring Steve Coogan, who plays the Factory Records boss Tony Wilson. They had even recreated the Hacienda night club for the shoot. It was identical to how I remembered it.

The film covered the history of the label and Wilson from its embryonic beginnings in a run-down club in one of the most deprived areas in Manchester, Hulme.

Joy Division, New Order, John The Postman, Duritti Collum, Happy Mondays, Pete Saville, The Smiths, Simply Red, and Mark E Smith from the Fall, were all referenced and referred to in varying degrees.

I was only on the periphery to the primary movers and shakers of Factory as the drummer with Lavolta, but watching the film was inspiring. All were larger than life characters. Some like Ian Curtis, Joy Division and New Order's manager Rob Gretton and producer Martin Hannett were no longer with us. However, all had a story to tell. Individually and collectively they left an indelible mark on the history of Manchester and on the worldwide music scene as a whole.

This had been my life before becoming involved in the world of aviation. It was vibrant, exciting and, looking back; life-affirming.

Yet, here I was, three decades later going head to head with individuals who would not have lasted five minutes in my previous world. Anyone who attempted a stitch-up like Scadeng, Hutchings, Thorington and Burke appeared to be attempting, would very soon be making the acquaintance of a baseball bat or experience what it was like to be hung out of a third-floor window by their legs. The thing was, for some reason, this kind of *'persuasion'* did seem to work. Actually, it worked pretty well.

I witnessed an example of this kind of 'illegal redress' when visiting Ash. Ash was my bandmate at the time. He was the guitarist in Lavolta. When I arrived at the flat, Ian and Reni from the Stone Roses were also there. It was late when Ian and Reni announced they were 'off to do some painting'. I remembered thinking to myself that it was a bit late to think about decorating your flat, but, each to their own.

Also in the flat was Paul who roadied for both Lavolta and New Order. Everyone knew him by his nickname 'Slim', and everyone loved Slim. A larger than life character, a genuine diamond, indeed, he still is. Slim lived next door to Ash in the horrendous 1960s designed Hulme Crescents.

After Ian and Reni had left to commence some DIY, Slim recounted what happened to him a few days previously.

He had returned that day from a world tour with New Order and was visiting Ash when they heard a noise from next door. It was someone ransacking his flat. The burglars must have thought he was still abroad. The occupants of the flat on the other side had heard the noise earlier and had already called the police. By the time Slim and Ash had run outside the police were there. Slim explained who he was and that it was his flat. The police told him and Ash to wait outside while they went in. Both expected the burglar to be marched out in handcuffs, but instead, a couple of minutes later, to their surprise, the police came out empty-handed. Had he escaped?

'If you guys want to go in and have a *little word* with him, we can wait out here.'

Funnily enough, Slim's flat never got burgled again. Ever.

That weekend I also found out what Reni and Ian had meant by painting. It seemed every major building in the city centre had 'Stone Roses' sprayed on it. This self-promotion must have worked. The band went from supporting Lavolta and B.F.G. to headlining Spike Island in front of thirty thousand fans.

Unfortunately, the option of hanging airline chief pilots out of windows was somewhat frowned upon in aviation circles. Indeed, I don't think I would have the strength to keep hold of Scadeng even if I'd wanted to.

I would have to come up with a more mature way of dealing with the issue, but one no less damaging, and no less effective.

It was now approaching three weeks since I had expected to be dismissed on the spot, yet, here we were, still nothing. Had the disciplinary letter joined the reply to my email of the 7th May, Scadeng's interview with Burke, the flight plans and the minutes to the disciplinary meeting into some perverse, really naff Thomas Cook version of the Bermuda triangle?

The longer the delay, the more I was convinced they would bottle it.

Had the *'by reasonable means'* comment I made during the disciplinary put the dampeners on Scadeng's and Hutchings' ramshackle intentions? I pictured Scadeng stamping his feet in the office 'I want, I want', as his HR carers informed him 'he couldn't, he couldn't'.

So, what were his options?

The first was to dismiss me. This was my preferred outcome. A nice clean, no holds barred attack on the entire company and individuals would follow. Not only was I convinced I would win, but I knew it would also feel fantastic. I would attack them with the legal equivalent of the baseball bat.

Their second option would be to demote me. Again, a reasonably easy situation to resolve. I would resign, claim constructive dismissal and still swing the bat for a legal home run.

My least preferred option would be for them to drop the case altogether claiming that new information had come to light. That would take the wind out of my sails and would mean I was caught between a rock and a hard place. Nothing would have changed, and the same individuals, who in my view were a threat to flight safety, would still be in charge. Fortunately, I also considered that to be the least likely outcome given the mentality of the antagonists.

Just as the film finished, with Coogan playing Wilson playing God, the phone rang. It was one of the most senior training captains in the airline. It was the same individual who had given me the heads up about the cosy relationship between the CAA flight ops inspectors and the senior management at Thomas Cook.

He was one of the good guys.

He said he was ringing to tell me about a strange conversation he had just had with Thomas Cook's chief training captain, Tim Cheal. They had been discussing several ongoing training issues when, without warning, Cheal suddenly switched the topic of the conversation. Unprompted, Cheal announced that the company had approached him to chair the disciplinary meeting. Now this made complete sense. It appeared that when it became apparent to Thomas Cook that I was aware of their plan to have Scadeng chair the meeting, the company had panicked. They had needed Cheal as the fall guy. Had things had become too toxic for Scadeng? The case was not the walk in the park he thought it would be. They now wanted Cheal to take any bullets flying around. According to the training captain, Cheal dodged the projectiles by telling HR directly, 'I'm not going anywhere near this'.

Cheal knew the damage that being involved in this case would do to his own reputation. More importantly, Cheal knew me reasonably well. He was fully aware I was not someone who would break the rules, or allow myself to be manipulated, nor did I suffer fools. Understandably he wanted to stay well clear of any involvement. He knew it would come at a personal cost. He knew I would make it personal.

I thanked the training captain for letting me know. I then surprised him when I let him into a little secret. I already knew. My mole in the Thomas Cook office had already told me that Cheal had announced that, indeed, he would not go 'anywhere near this'.

A few days later, a full eighteen days after the meeting, I finally received the letter from Scadeng with the outcome of the disciplinary. Still no minutes from the meeting, still no answer to my email.

According to Scadeng, my actions "raise very serious concern regarding integrity and the trust and confidence that the company must have in its commanders."

So, why were they so concerned about my integrity?

The next paragraph explained.

"On the balance of probabilities, I believe that as you first raised fatigue only some hours later on being asked to call the company after landing, that you were *dishonest in asserting that the reason for refusing the duty of the 7th May 2014 was the onset of fatigue on the 6th."*

Well, what do you know and who would have guessed? Remember, Scadeng was the individual who changed his vote after the alleged curry meeting. Scadeng was the individual who, to my mind and several colleagues, had sold the BALPA membership down the river with regards to the share issue. Scadeng was the individual who jumped from being the chairman of the BALPA CC to chief pilot without any cooling-off period.

Now, here he was, in a legal document, claiming I was "asserting that the reason for refusing the duty of the 7th May 2014 was the onset of fatigue on the 6th."

There it was, in black and white, a lie I knew he had to make if he was not going to lose face in front of Hutchings and the entire company.

When I read that, my honest, immediate and genuine reaction was 'thank you for being such a complete and utter tool'.

Scadeng had cleared the way for me to do something that I had been aching to do for months. It would allow me to give him and his mates the biggest legal kicking they would ever get, and all without resorting to physical violence.

He had done precisely what I was hoping he would do. He had done what I had corralled him to do.

Scadeng had been forced to resort to falsifying what was said in total contradiction to every single piece of evidence.

I knew he would. He was incapable of backing down. It had become his only option.

Did Thomas Cook legal have any idea what an absolute open goal this letter was for BALPA legal and myself? If they didn't, then, in my view, they were as incompetent as our moist friend.

Scadeng had even gone as far as to use the term that I had coined.

*'Onset of fatigue.'*

However, for his case not to collapse, he had to claim that I had said I was suffering the onset of fatigue on the 6th.

The only time I had actually used that phrase was the morning of the 7th. The email to Scadeng, written and sent on the 7th confirmed that fact:

"For the first time in my flying career, today I rang up crewing to inform them that I would be unable to perform a flight duty, not due to illness, but due to my very real concern that I was experiencing the onset of fatigue which would erode flight safety if I completed today's duty."

So, what did Scadeng intend to do about it?

"This raises very serious concern regarding integrity and the trust and confidence that the company must have in its commanders. With this in mind, I find that the appropriate sanction would be dismissal."

Perfect, it was time to dust off the metaphorical baseball bat. Oh. Hold on. Scadeng continued.

"However, due to the mitigation arguments put forward on your behalf by your BALPA representative and your length of service, I have exceptionally decided to reduce the sanction to a final written warning to remain current for a period of two years."

Damn.

I am now confused. Scadeng claimed that I had no integrity and that the company had serious concerns in their trust and confidence in me as a commander. Surely if that was the case, that would justify immediate dismissal? If he did genuinely hold those views, then the very least Scadeng should do is demote me. How could he possibly justify someone he didn't believe had integrity, or he did not trust, command a Thomas Cook aircraft with hundreds of passengers onboard? Any reasonable person would consider this to be a dereliction of Scadeng's and the airline's duty to keep its passengers safe.

He continued,

"During the disciplinary process, you showed no insight into your misconduct in seeking to avoid a sequence of duties you did not want by seeking to rely on reasons that were not genuine. For you to remain employed as a captain, it is vital that you make best efforts to repair the damage to the relationship of trust that has been caused by your dishonesty, and display the standard of behaviour and decision making that is expected of a Thomas Cook Airlines captain. That process might start with an acceptance of this outcome and a determination to put this event in the past and move on positively."

Scadeng was not just using a spade to dig his own hole. He had brought in a JCB.

There was only one logical conclusion one could come to regarding Scadeng's comment that I had not displayed '*the standard of behaviour and decision making that is expected of a Thomas Cook Airlines captain*' .

There could be no other way to explain it. Thomas Cook management considered that the standard of behaviour expected would be the willingness of its captains to disregard legislation and passenger safety and instead, comply with any falsified documentation.

The main issue of this letter was that, to make the above determinations, Scadeng needed evidence. His problem was, he had none - zilch. So what could he do? The only thing he could do. Make it up, falsify it.

You had to admire their tactics, though. It was a genius move. Well, so they thought.

I hadn't been dismissed nor demoted. I, therefore, had not suffered any reduction in salary.

They had instead given me a two year written warning. A slap on the wrist that they assumed I would accept while being thankful I still had a job.

It would mean they had avoided an external legal challenge to their behaviour and the dire PR that would result in a public case.

They no doubt considered a two-year final written warning was a masterstroke.

During those two years, with a final written warning over my head, Scadeng and Hutchings could wait and pick me up on any minor misdemeanour not related to health and safety and dismiss me. As the final part of the letter stated:

"Please be aware that for the duration of this warning, if your conduct falls below the required standard regarding this or in any other matter, then further disciplinary action may be taken, which could result in your dismissal."

Hutchings and Scadeng undoubtedly considered it was now all wrapped up. They were, however, about to get a very rude awakening.

I suspected Thomas Cook legal, HR and managers had yet to make the acquaintance of Ms, Jennie Fecitt.

It was time to take the gloves off…

## *Acnowledgements*

Without doubt, writing the three volumes that comprise the series Pulling Wings From Butterflies has been the most intense and mentally demanding project I have ever undertaken. It has been made infinitely easier with the support and love of my beautiful wife Lisa, and my amazing children, daughter Katie and son Jake. You are the most important people in my life.

Heartfelt thanks to my old mucker from the Saab days, Ian 'Baldrick' Kenyon, for wading through page after page, correcting my sometimes 'artistic' interpretation of the Queen's English. You sir, are a diamond.

Likewise thank you to Bea Stevens for catching those pesky errors that got through the net and Jake for applying the final polish.

Many thanks to the mANAGER Hugh Chambers for being my touchstone and playing devil's advocate when I needed a different perspective.

To Dr Rob Hunter, Ian White, Capt. Pete Elliot and Capt. Richard Minashi, thank your believing in me when others turned away and listened when no one else would or wanted you to. Also, thanks to Claire Coombes for her support, a young lady wise beyond her years.

Thanks to Capt. Mark Priestley for making the initial approach to the legal titan that is Seamus Sweeney who graciously confirmed my course was true.

To all my past flying colleagues, thank you for the memories and fun. Notably, thank you to Alex Lawrence for the silliness, support and professionalism at the appropriate times (did I tell you about BMI Alex?) and to Chas Martin for confirming my suspicions and for checking in regularly with me along the way.

I will forever be grateful to Capt. Stefan Ratayak for showing me the value of *spare capacity* and sorting out my steep turns along with all the training captains who allowed me to access their years of experience, notably Capt. Simon Spence, Capt. Pete Elliot, Capt. Ian Goodwin, Capt. John Ward, Capt. Victor Peirce and Capt. John 'Bloody Hell Man' Smith.

Massive thanks to Judge George King for the pearls of wisdom and updates on the latest post-punk releases and to Markus Beldig for the regular rambling conversations.

To my B.F.G. brothers, Barry 'Bozstar' Williams, Zero, Dan Cunningham, Elliot Wheeler, Andy Wright and the guy who was my partner in crime Keith Lambert right at the start...Love you guys and love the racket we still make.

I'm grateful to the B.F.G. fans who have stuck with us over the last thirty years and made our return such a fantastic and memorable experience and to Jürgen Jakob for coming up with the ridiculous idea that B.F.G. should reform in the first place.

Thanks to Pete Hook for giving B.F.G. a leg up when we needed it and Mick Mercer for helping open the door. Thanks also to Steve Craig, Paul ‘Baggy Batheram, Joe Cassidy, and Elaine Thomas, for being a big part of the B.F.G. story and Guy Hatton, Andy Clegg, Andy Whitaker, Paul Wayman, Andy Fletcher, Graham ‘Acky’ Atkinson and Steve O'Donnell for coming along on the ride.

Thanks to John Robb, Peter Byrchmore, Nick Brown and Phill Newall for inviting Amelia Chain and myself to join the Membranes on the amazing Mark Lanegan European tour. It was exactly what was needed, exactly when it was needed. To tour mates Jeff Fielder, Aldo Struyf, Christoph Claeys, Frederic Lyenn Jacques, Roy Akkeren, Paul Delpeut, Sal Priem and of course Shelley and Mark Lanegan himself, a most profound thanks for making it such an unforgettable way to bring life full circle.

To all those new friends made during those astonishing six weeks, I hope to see you all again in the near future.

Made in the USA
Monee, IL
04 July 2026

56401619R00144